Field Manual for Global Low-Cost

Water Quality Monitoring

Second Edition

William B. Stapp, Ph.D.
Mark K. Mitchell, M.S.

Contributors:

Mike Appel, Ethan Bright, Mare Cromwell, Timothy Donahue, Cheryl Koenig, Heike Mainhardt, Susan Maruca, Mary Mitsos, Shelley Price, Keith Wheeler

Editing by:

Rachael Cohen

KENDALL/HUNT PUBLISHING COMPANY
4050 Westmark Drive   Dubuque, Iowa   52002

All photos by William B. Stapp unless otherwise credited.
Cover photos by William B. Stapp.

This manual is used in many national and international watershed programs linked through the Global Rivers Environmental Education
Network (GREEN). For further information regarding this international
network contact:

GREEN
206 South Fifth Avenue, Suite 150
Ann Arbor, Michigan 48104, USA
Telephone: 313-761-8142
FAX: 313-761-4951
Internet: <green@green.org>
World Wide Web: <http://www.igc.apc.org/green>

| Dedication

To the many international educators who have committed expertise, time and energy towards the development of low-cost water quality monitoring techniques so that students can monitor local waters and take appropriate actions on their findings. It is due to the insight and caring nature of those noted above that this manual has been prepared.

Contents

| Preface

We want to express our sincere gratitude to the many people that have brought to GREEN's attention the need to produce this *Field Manual for Global Low-Cost Water Quality Monitoring*. In particular, we want to recognize the work and insight of Rob O'Donoghue, Natal Parks Board, South Africa; Patricia Bytnar Perfetti, University of Tennessee, Chattanooga, Tennessee; Charles Terrell, National Water Quality Specialist, Soil Conservation Service, Washington, D.C.; Esther de Lange, International Water Tribunal, Netherlands; Sue and Col Lennox and the National Waterwatch Program, Australia; and Mike Appel, David Bones, Mare Cromwell, Heike Mainhardt, and Keith Wheeler of the Global Rivers Environmental Education Network (GREEN).

We want to acknowledge the entire GREEN staff for their enduring commitment to produce a low-cost water quality monitoring manual, so that youth throughout the world can monitor local waters and work with citizens and officials in obtaining a more healthy environment for all people.

In particular, we want to recognize the following professionals for reviewing and providing valuable comments on this manual: Dr. David Allan, Dr. Jonathan Bulkley, Dr. Paul Nowak, Dr. Michael Wiley, Mr. Tim Donahue, Mr. Steve Hulbert, Dr. Paul M. Kotila, Ms. Lisa LaRocque, Ms. Lisa Bryce-Lewis, Mr. Peter Oliver, Mr. Larry Price, Mr. Joseph Rathbun, Mr. David Schmidt, Mr. Keith Wheeler, and the Australian National Waterwatch Program. It is important to recognize the many private and public bodies that have provided valuable ideas and funding that have enabled GREEN to produce this manual, such as: The General Motors Corporation; United Nations Environment Program; United Nations Educational, Scientific and Cultural Organization; United States Environmental Protection Agency; National Science Foundation; United States Department of Education; National Consortium for Environmental Education and Training, The University of Michigan; Public Interest Research Group in Michigan; Eisenhower and TERC Funds; Walpole Island Native American Heritage Center; Michigan Department of Natural Resources; Dow Chemical of Canada; The George Gund Foundation; Frank Butt Memorial Foundation, Australia; and the International Polysar Corporation.

The authors would like to recognize the talent of Shelley Downes who prepared the graphic artwork on benthic macroinvertebrates in Chapter 6. Shelley was a high school student who participated in the Rouge River Water Quality Monitoring Program. Her extraordinary ability was called to our attention when she was fourteen, and she has prepared drawings for several GREEN publications. It is also an honor to recognize Mr. John Henry for his computer art work displayed in several chapters.

It is with deep respect that we note the creative talent, skills and personal commitment of Rachael Cohen and Kevin Bixby for their editing and layout. Both Rachael and Kevin contributed an inordinate amount to time and energy to make this manual clear and readable for our global audience.

Finally, we would like to express our sincere gratitude to Gloria and Tara, and other members of our families, for the commitment and support they have provided through the years to enable this publication to be produced.

Purpose of the Manual

Introduction

Water quality monitoring has often been reserved for those with access to significant funding and advanced technical training. There are, however, many approaches to measuring water quality and assessing the quality of a catchment that require little or no funding and only a few well designed workshops. Throughout the world, access to lower-cost monitoring approaches can lead to an understanding of not only water quality problems, but the sources of these problems. This understanding helps those people most affected by water quality problems to evaluate and change the situation.

Purpose

This manual is designed to complement the *Field Manual for Water Quality Monitoring, 11th Edition* (Mitchell and Stapp, 1997). The purpose of the *Global Low Cost Manual* is to provide monitoring options that are inexpensive, require little or no technology, and provide measurements and observations valuable to a better understanding of the condition of a river and its catchment.

The Global Rivers Environmental Education Network (GREEN) assists catchment monitoring programs that reflect the entire resource spectrum—from programs having adequate funding to programs having few funds. Although this manual can be a resource for any program along this spectrum, it has been designed in response to the needs of less well-endowed programs.

Watershed or Catchment?

The term catchment is used instead of watershed in this manual for two reasons: *catchment* seems more descriptive of the functionality of an area of land that "catches" precipitation and overland surface flow and sends it to a common channel; and, *catchment* is used more widely at the international level. See the Glossary (Appendix C) for definitions of this and other terms used in this manual.

Figure 1.1. GREEN staff discussing the involvement of the government, community, and schools in a water quality monitoring program with Taiwan officials.

The GREEN international office is based in the United States (Ann Arbor, Michigan) but it is a network with strong international participation—with 46 country coordinators. This manual is a compilation of assessment work that has been carried out by scientists, schools, government agencies, and non-governmental organizations from the United States, Africa, The Netherlands, Australia, India, and many other countries. Readers, particularly international readers, should view this manual as a work in progress.

GREEN may not fully know the wealth of assessment approaches currently available. Accordingly, GREEN seeks to continually improve upon the assessment approaches currently used, as well as discover and share new approaches internationally.

Design of the Manual

The focus of this manual is to provide descriptions of monitoring approaches, and also activities that help readers understand key concepts and build important skills. It is designed to be a practical manual that offers equipment-making instructions, support materials, and handouts. The emphasis is upon quantitative collection of water quality data that readers can use to better understand a catchment and to act to solve water quality problems. The manual has been designed to provide background information about the biological, physical, and chemical nature of rivers and their

Figure 1.2. Use of terracing to grow rice and to protect the hillside from major soil erosion in Nepal.

catchments, and considers the human dimensions of water quality as reflected in land use practices. It offers monitoring options appropriate for those with little or no funding, as well as for those who have more funding available to support monitoring efforts.

How to Use This Manual

Readers unfamiliar with GREEN will want to read Chapter 2 to gain an understanding for the international efforts that have led to this manual. Chapters 3 and 4 focus on rivers of the world and the human activities that impact catchments and river systems. The study design chapter (Chapter 5) may be helpful to those readers who wish to help coordinate or design a water quality study. Other readers may simply be looking for an introduction to benthic macroinvertebrates and monitoring (Chapter 6). Those using the manual in an educational setting may want to pay special attention to Chapter 7, which offers activities designed around key concepts and measurements. Chapter 8 offers three levels of conducting physical and chemical tests—from purchasing chemical test kits to making your own testing equipment, and provides a set of activities for carrying out physical-chemical testing of an aquatic system. Chapter 9, about action-taking, is helpful for readers who have completed some monitoring and need guidance in setting up a framework for implementing change at an individual, school or community level.

Figure 1.3. Egyptian irrigation canals used for domestic chores pose serious health problems.

Overview of this Manual

Each chapter is the product of the catchment monitoring models that GREEN has developed internationally. They are a reflection of much thought and effort by the educators, citizens, scientists, and community and government leaders who comprise the GREEN network.

This manual is organized as follows:

- Chapter 1, *Purpose of the Manual* offers an introduction to the *Field Manual for Global Low Cost Water Quality Monitoring*, including the importance of low-cost monitoring, purpose of the manual, design of the manual, use of the manual, and an overview.

- Chapter 2, *GREEN Experiences,* describes the educational philosophy and world view underlying the Global Rivers Environmental Education Network. This chapter relates the history of GREEN as a model, organization, and network; describes the international dimensions of GREEN; outlines the major elements of GREEN; provides a scope and sequence approach to a catchment curriculum for primary and secondary schools; and lists components of the GREEN network.

- Chapter 3, *Land, Rivers, People,* discusses the interconnections between water quality and the human dimensions within a catchment. Specifically, land use issues such as urbanization, salinization, and unwise agricultural and forestry practices are described. Waterway alterations, including dams and impoundments, channelization, wet-land drainage, and dredging are also covered. Descriptions and examples of point and nonpoint source pollution are given. The issues raised by this chapter establish a clear need to assess the condition of a river and its catchment, the importance of public participation, and the need for widely available and accessible approaches to assessment.

- Chapter 4, *Rivers of the World*, provides a broad overview of rivers of the world and how their catchments are shaped by soils, vegetation, climate, underlying geology, and other factors. This chapter covers the unequal distribution of water around the world and looks at the impact of discharge and sediments on river systems. These factors are related to the biodiversity and global distribution of fish and other aquatic life.

- Chapter 5, *Study Design*, examines several study designs that illustrate the interconnections among assessment areas described in later chapters, and gives the reader a guide for their application. Elements of study design include evaluations of the needs or questions being asked, the skill level of those participating, and the level of funding that can be used for equipment, data analysis, and interpretation. This chapter also provides a guide for safe sampling and measurement of rivers.

- Chapter 6, *Benthic Macroinvertebrates,* offers an introduction to benthic macroinvertebrates and to biomonitoring using macroinvertebrates. The behavioral and physical characteristics of benthic macroinvertebrates, the use of macroinvertebrates as indicators of water quality, a guide on where and when to sample, the analysis of benthic indices, and a how-to section on benthic sampling devices and equipment are included.

- Chapter 7, *Catchment Analysis Activities,* provides a range of activities: catchment activities (satellites, mapping, and land use), assessing the riparian and bank areas of the river, the phytoplankton and macrophytic communities, and indices and techniques for measuring the macroinvertebrate community. Important background information on these monitoring areas is also provided.

- Chapter 8, *Physical and Chemical Water Quality Monitoring Activities,* details low-cost physical and chemical monitoring analysis activities. Three options for testing are described: Option One, purchasing simple ready-made commercial water quality kits; Option Two, purchasing refill chemicals to run water quality tests, and; Option Three, very low-cost commercial kits, and obtaining materials locally to produce school-made equipment.

Figure 1.4. Home-made water quality monitoring equipment to test for turbidity in local rivers.

- Chapter 9, *Action-Taking and Activities*, considers action-taking Activities. What do we do with the data collected in a water quality monitoring project? What is action-taking? What forms can action-taking assume? This chapter begins with a brief overview of research underlying some action-taking approaches, followed by several action-taking activities. These approaches are illustrated by three case studies demonstrating action-taking—two from the Rouge River (United States) and one from Australia.

- Appendix A, *Availability and Accuracy of Commercial Kits and Equipment*, provides important guidance on commercial kits, and for constructing homemade kits.

- Appendix B, *Handouts, Data Sheets, and Surveys,* these handouts and data sheets are invaluable because they are often time-consuming to develop from scratch. They may be removed from the manual and copied for use with each activity.
- Appendix C, *Glossary,* contains definitions of terms used throughout the manual.
- Appendix D, *Bibliography,* listing sources used in compiling this manual, as well as suggestions for further reading.
- Appendix E, *Index,* a useful reference section providing the page numbers within this manual on which important terms and topics can be found.

CHAPTER

GREEN Experiences

Introduction

"Globalization" refers to the increasingly interdependent world, linked by a closely coupled world economy. This shrinking world is brought closer together by massive environmental problems and issues that transcend national and even continental boundaries—issues that can be addressed only through an unprecedented degree of global cooperation.

One major challenge that will increasingly confront environmental educators is to develop curricula and instructional strategies that emphasize the global aspect of local environmental issues but do not overwhelm the students or cause them to lose hope. How can we educate and empower students to take action on local issues, while simultaneously developing within them a global, cross-cultural perspective? How can we best encourage this first generation of truly planetary citizens to assume responsibility for their shared home?

One promising approach to meeting this challenge is being taken by the Global Rivers Environmental Education Network (GREEN), an international network that seeks to bring students and teachers, and their communities around the world closer together through the bond of studying and improving our common river systems. GREEN was initiated by professors and graduate students of the University of Michigan's School of Natural Resources and Environment in 1989, and has developed into a multifaceted, global communication system that invites participants to examine ways that land and water usage and cultural patterns influence river systems, and vice versa. The Network encourages learners to become involved in complex, real-world concerns that extend across traditional boundaries.

GREEN works to achieve three interrelated goals:

➤ To acquaint students with the environmental problems and characteristics of their local catchment, giving them "hands-on" experience in chemical, biological, and sociological research.

➤ To empower students through community problem-solving strategies, enabling them to see the relevance of subjects they study in school to the "real world."

➤ To promote intercultural communication and understanding, in order to foster awareness of the global context of local environmental issues and the significance of cultural perspectives in choosing effective problem-solving strategies.

Why Rivers?

Rivers were chosen as the central focus of GREEN primarily because they are a reliable and informative index of the environmental quality of their catchments. Rivers also form a connection for relating chemistry to biology, and for relating the physical sciences to the social sciences and humanities. Rivers bind together the natural and human environment from the mountains to the sea, and from farmland to the inner city. In fact, 85 percent of the world's human population lives on or near a river. For these reasons, the study of rivers forms a coherent curricular framework for the study of a wide range of environmental issues. Rivers also provide an historical perspective on cultures and society, forming an ideal basis for learning about cultural diversity and engaging in cross-cultural dialogue.

│ Figure 2.1. Families involved in "driving" native fish into a net in Benin, Africa.

Through involvement in a network on local rivers, students share information, techniques, and different approaches to problem-solving. They also learn that their investigations are valued by their peers elsewhere in the world. Students are motivated to further their understanding of their catchments and to work to resolve some of the water quality problems they have discovered. GREEN is thus a program designed to bring individuals closer together and to encourage them to develop a sense of responsibility for their communities and their planet simultaneously.

The History of GREEN and Global Activities

Under the guidance of William B. Stapp, a committee of 26 university students with backgrounds in environmental education and international issues was organized as an advanced level Environmental Education class in January of 1989. The committee members established the vision and goals of GREEN as they laid the foundation for this international network. The committee organized and facilitated 22 workshops in 18 nations in Africa, Latin America, Europe, Asia (including the Middle East), and Australia during the summer of 1989. The workshops brought together educators, administrators, students, citizens, resource specialists, and representatives from governmental and nongovernmental organizations to exchange ideas on catchment programs. One of the aims of the workshops was to discuss approaches to experiential, interdisciplinary environmental education, and to explore how water study programs could support the educational goals of each nation.

The Creation of the GREEN Infrastructure and Networking

Many of the nations that hosted or attended workshops during the summer of 1989 have now established school-based water monitoring programs. For example, Taiwan initiated two pilot programs that are on-going. Schools in Germany created an environmental monitoring network. Israel has incorporated river studies into their national senior high school curriculum and established a water monitoring program on the Na'aman River involving Arab-Israeli and Jewish-Israeli schools. Many countries have appointed GREEN country coordinators to oversee GREEN activities and prepare articles and other educational materials. In addition, funding from private and public sectors has been allocated to develop programs and obtain equipment and supplies.

Figure 2.2. University students and professors in India mix and distribute chemicals to secondary schools for monitoring local rivers in Ahmadabad.

With so much interest shown in learning more about student water monitoring and networking, in 1989 GREEN focused on creating a globally accessible network and on the development and dissemination of relevant resources to participants. GREEN began to publish a newsletter, develop curriculum guides for various age levels, conduct teacher training workshops, and support other national and international river education projects.

The Expansion of GREEN

In 1993, GREEN became a private, not-for-profit organization whose mission is to improve education through a global network that promotes catchment stewardship. GREEN's strategies for achieving this goal include providing program support, enhancing information exchange, developing cross-cultural opportunities, encouraging educational reform, and conducting educational research.

Worldwide response to GREEN has been phenomenal. GREEN has grown to reach thousands of students involved in catchment projects in countries as widespread as Bangladesh, the Czech Republic, and Argentina. There are now GREEN Country Coordinators in 46 nations and active programs on all continents. Some nations, such as Germany and Australia, have established extensive national networks to facilitate communication.

Within the U.S., all states have developed catchment-wide school programs and many other schools are monitoring independently. The recent "Teacher Enhancement Program," funded by the National Science Foundation, allows teachers in five catchment programs in the United States to work together to further develop the GREEN catchment education model and to create plans to disseminate the model to other interested schools in their regions. GREEN is also working with Native American educators to help assess their environmental education needs and provide program support.

The development of GREEN can be largely attributed to a sound environmental education model and a vision transcending cultural boundaries with global environmental issues. The committed work of GREEN staff and University of Michigan students and faculty have propelled the project to realization, this complemented by the international community of dedicated environmental education leaders who participate in GREEN.

Key Ideas for GREEN's Global Activities

A number of ideas and concerns raised during the original international workshops in 1989 continue to inform the development of GREEN's activities.

➤ *Practical concerns about water quality programs.* There was great interest among the participants of the 1989 workshops in water monitoring programs that enabled students to link education to real-life experiences, work between disciplines, share information through computer networking, and take action on the information collected. Some participants expressed concerns that science education should remain value-free, that leaving school grounds would not be permitted, and that water monitoring kits and computers were too costly.

➤ *Water quality varies.* The workshop participants observed rivers that varied in quality from very pristine to highly polluted. The most polluted river contained at mid-day: 0.0 ppm dissolved oxygen, 1.2 million colonies of fecal coliform per 100 ml of water, 138 ppm biochemical oxygen demand, and high concentrations of heavy metals and toxic organics. Nitrogen levels in some estuaries had increased by 200 percent since the 1950s. One bay has received 600 tons of inorganic mercury since 1953.

➤ *Water quality influences health.* Eighty percent of sicknesses in the world are due to unsafe water and poor sanitation. People in the Southern hemisphere are deeply aware of the linkage between water pollution and health issues. Over 4,000,000 children below the age of five from

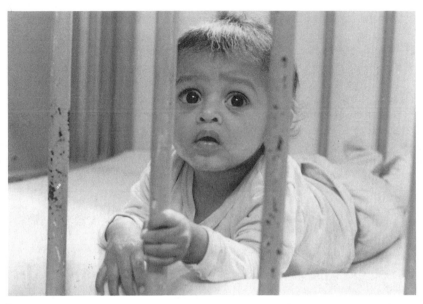

Figure 2.3. Four million children under the age of five die each year in the Third World from contaminated water.

the Third World die each year directly from water pollution. This figure is larger than all deaths, at all age levels, from all causes in Europe and North America combined. In the developed world, diarrhea from impure water is uncommon, but in the developing world it kills more people than cancer. While the Rio Conference in Brazil, 1992 emphasized global warming, ozone depletion, and threatened species, little attention was placed on the 7.8 million poor people who die each year in the Third World from what they breathe and drink. In many parts of the world, there are no pipes to distribute clean water to residents nor waste water treatment facilities to reduce pathogens from entering the waterways.

➤ *Involving indigenous people.* Water quality monitoring programs in many regions of the world are striving to involve indigenous people in an effort to incorporate a broadening perspective of environmental sustainability. These programs emphasize the close collaboration of elders and the collection of oral histories and community-based learning.

➤ *Development of lifelong observational skills.* It was evident during GREEN's trips to Africa, Latin America, and Asia that water quality monitoring should emphasize low technology. African educators would sit on the river bank and share astute observations: rocks covered with algae indicating nutrient enrichment, the meandering river cutting the bank and depositing the soil on the flood plain, plant macrophytes indicating water temperature and pH, riffle areas contributing dissolved oxygen to the aquatic ecosystem, or small concentric rings in the water marking the site of emerging insects. African educators' observations were consistent with water quality data obtained with scientific water monitoring kits and the systematic macroinvertebrate assessments.

Figure 2.4. An Inuit woman in Greenland testing the local water supply for nitrogen.

➤ *Making low-cost monitoring equipment.* Many schools throughout the world lower expenses by making their own water quality monitoring equipment. These include collecting nets, extension rods, turbidimeter tubes, temperature gauges, and other materials.

➤ *Interest in using computers.* Some schools use computer networks to share information regarding their river monitoring programs, or to access national databases on rivers. These schools are interested in making better use of international computer networking opportunities. Teachers in some other schools viewed computers with less enthusiasm, believing that they would overshadow less high-tech activities and be more costly, time-consuming and impractical.

➤ *Use of aerial and satellite images.* Aerial and satellite images were found to be highly useful in monitoring catchments to determine current and changing landuse practices. Schools have found ways to reduce expenditures by obtaining permission to use photographs and to make photocopied reproductions for classroom use.

➤ *Cross-cultural opportunities.* Students and teachers in water quality projects were interested in exchanging personal and cultural perspectives, concerns for the environment, and ideas for improving the catchment with others in similar projects around the world. Students communicate by mail and computer to share ideas and actions taken. The goals of cultural exchange programs are to stimulate greater international awareness in students and motivate students to develop concern for improving their local waterways.

➤ *School structure is a factor.* School curricula vary greatly throughout the world. Some nations are very decentralized and flexible in permitting interdisciplinary water quality programs, while other nations maintain highly centralized systems where students are prepared for rigorous national examinations. However, one such nation permits students to substitute an independent project, such as an individual river study, for the national biology exam.

The experiences of GREEN participants since 1989 have brought to light the following characteristics of successful programs:

➤ *Catchment Orientation*—identification of boundaries and geography of one's catchment; focus on the entire catchment; networking within and between catchments.

Figure 2.5. Native American elders using school computers to communicate with other Canadian bands in the Great Lakes.

➤ *History and Culture*—knowledge of indigenous settlements and cultural and environmental relationships within the catchment; early immigration patterns; the process of urbanization; municipal sanitation concerns and efforts.

➤ *Land-Use Practices*—provision for field studies and excursions; use of maps and aerial images to note changing land-use practices; identification and recognition of the value of wetland areas.

➤ *Benthic Organism Studies*—noting the value of biological monitoring; identifying the diversity of aquatic life in the river; studying what organisms are present or lacking.

➤ *Nine Water Quality Tests*—testing the water for some of the physical or chemical parameters; the making or use of low-cost monitoring equipment.

➤ *Computers and Networking*—use of graphing paper or computers to analyze river data; communication within the catchment in reference to the data collected and analyzed.

➤ *Laws and Regulations*—identification of the laws and regulations that govern water quality in the region; familiarity with regulatory agencies in the catchment or region.

➤ *Cross-cultural Exchange*—development of a cross-cultural partner within one's catchment or in another region or nation.

Figure 2.6. Water from a local river in Bombay, India is directed to land and used by laundry washers and then returned to the river system.

➤ *Student Congress*—discussion of water quality data and the sharing of concerns with others in the catchment; presentation of water quality data and concerns to the regulatory agencies in one's region.

➤ *Futurism*—considering the consequences of environmental trends that are shaping our future (forest loss, soil erosion, water scarcity, coral reef destruction, etc.); focus on environmental sustainability concepts.

➤ *Action Taking*—taking appropriate action on the environmental information collected; identifying examples of successful student action.

➤ *Evaluation*—development of an on-going evaluation procedure; determining what changes are occurring at the student, teacher, institution, and/or community level.

Figure 2.7. Students at Aikahi Elementary School in Kailua, Hawaii are working with the community to help reduce sediments entering coastal waters.

➤ *Integrated catchment curriculum*—Some schools have developed "Catchment Curriculum Programs," that start in the primary grades and follow a scope and sequence approach culminating at the secondary levels. An example of one such program follows (Walpole Native American Program, Walpole Island, Canada):

Grades 1-2
    History and Culture (settlement, cultural patterns)
    Geography Concepts (precipitation and drainage)
    Wetlands (defining wetlands, value of wetlands)
    Aquatic Systems (sealed world, recycling)
    Benthic Organisms (aquarium study, food chain)
    Water Quality Tests (temperature, aquatic plants)
    Multi-Media Learning (video tapes of the environment)
    Pollution (definition, danger of pollutants)
    Computer (games, word processing)
    Empowerment (personal responsibility)
    Congress (share findings, student concerns)

Grades 3–5
    History and Culture (historical patterns, oral history)
    Geography Concepts (mapping and land usage)
    Wetlands (endangered and threatened species)
    Aquatic Systems (food web, energy cycles)

Field Trip (river sensory walk, observations)
Benthic Organisms (sequential comparison index, plants)
Water Quality Tests (DO, pH, BOD, Turbidity)
Pollution (kinds of pollutants—sediments and toxins)
Multi-Media Learning (audio cassette/recordings)
Community Interviews (citizens, policy makers)
Computer (record information, spread sheet of data)
Empowerment (school responsibility)
Congress (present data, share concerns/actions)

Grades 6–8

History and Culture (nature and traditional art)
Geography Concepts (model building and natural resources)
Wetlands (plant and animal extinction)
Aquatic Systems (cycles, disturbances)
Benthic Organisms (pollution tolerance index, analysis)
Pollution (thermal, point and non-point)
Water Quality Tests (nine water quality tests, WQI)
Survey (home hazardous survey, school survey)
Field Trip (landfill, waste water treatment plant)
Multi-Media Learning (portable computers/field data)
Computer (graphing data, analyzing data)
Cross-cultural (partner catchment-national)
Empowerment (neighborhood responsibility)
Futurism (environmental ecological balance)
Congress (share results, develop recommendations)

Grades 9–12

History and Culture (archaeological excavation)
Geography Concepts (satellite images, ground truthing)
Wetlands (national and international policies)
Aquatic Systems (pollution, sustainability)
Benthic Organisms (mayflies, stoneflies, caddisflies)
Water Quality Tests (nine tests, WQI)
Pollution (heavy metals, bioassay testing)
Laws and Regulations (past, present, future)
Cross-cultural (partner catchment-international)
Multi-Media Learning (zap shot camera/CD ROM)
Computer (telecommunication)
Empowerment (community/regional responsibility)
Futurism (sustainable futures, eco-management)
Congress (visualization, political action)

National Planning Efforts in Australia

Australia has put together one of the most progressive and successful National Waterwatch Programs, with an allocation of $2.9 million over the initial three years. The Program established a National Waterwatch Office and Facilitator, Waterwatch Coordinators in every state, as well as Regional/Catchment Coordinators. The program also established a

Waterwatch Advisory Committee, National Waterwatch Steering Committee, State Waterwatch Steering Committees, and Community Groups representing a cross-section of landowners, business representatives, governmental officials, and the educational community. The process provides national planning, data base collection, general funding, cooperation, materials and equipment to carry out a well-planned water quality monitoring program throughout the nation. In one of the communities, Sydney, local residents passed legislation to tax themselves to improve water quality within the region. Part of the funding is used to support a highly successful water quality monitoring program, involving a network of over 200 high schools in the region.

Components of the GREEN Network

Using communication to affect environmental change and empower students is the backbone of GREEN's philosophy. The Network presently disseminates a quarterly newsletter to educators, governmental officials and other resource persons in over 135 countries. A series of GREEN International Computer Conferences have been established. GREEN has created opportunities for cross-cultural exchange between schools in different nations and also serves as a clearinghouse and a resource for catchment education programs.

➤ *The GREEN Newsletter*
The GREEN Newsletter is the most extensive and important communication tool of the Network due to the accessibility of the mail system internationally, and the strength of the written word to impart information and foster the GREEN spirit. Examples of newsletter article topics include: ideas for starting a water monitoring program, low-cost monitoring techniques, and models for student action-taking and community involvement. Each issue highlights exciting programs that serve as examples of local water quality education programs around the globe.

➤ *The GREEN International Computer Conferences*
The GREEN Conferences are international, electronic forums for the exchange of student-collected water quality data, reports from catchments, solicitation of cross-cultural partnerships, and the ideas and concerns of students, teachers and other professionals engaged in GREEN programs. Participants are able to communicate their experiences and receive almost instantaneous responses from around the world.

Figure 2.8. Students monitoring in a clean river for benthic macroinvertebrates with home-made equipment.

In addition, individual catchment programs use on-line networks to host local and regional computer conferences. These conferences allow the students to enter their data and communicate interactively with other schools in their catchment.

The GREEN International Computer Conferences, and many of the catchment-specific computer conferences, are hosted by the Association for Progressive Communication (APC) networks, an international coalition of independently operated computer networks in 18 countries that together extend their services to more than 135 countries. Via EcoNet in the United States and its APC counterparts, the GREEN conferences can be shared with regional, state, and local educational telecommunications networks around the world.

The GREEN office in Ann Arbor, Michigan, may be contacted for details via Internet at <green@green.org>. Information about GREEN is also available via Internet Gopher at <gopher.igc.apc.org> in the Education & You/Projects menu, or via World Wide Web at <http://www.econet.apc.org.green>.

➤ *The Cross Cultural Partners Watershed Program.*
The Cross Cultural Partners Watershed Program has sparked remarkable interest among schools worldwide. GREEN matches schools involved in catchment projects in different countries to enhance cross-cultural sharing. The pilot student-to-student links, initiated in 1991, consisted of schools in the United States, Canada, Mexico, Hungary,

Figure 2.9. One of the most successful school-based water quality monitoring programs in the world started in Sydney, Australia, and has expanded internationally.

Australia, New Zealand, and Taiwan. Since then, other partnerships have been established on all continents. Using local water quality issues as a springboard for discussion, students exchange personal cultural perspectives and concerns for the environment, along with ideas for improving their environment. Students generally communicate by mail and computer to share their thoughts.

The aim of this cultural exchange program is to stimulate greater international awareness in students while motivating them to work toward improving their local waterways. GREEN is developing materials that will help partnered schools achieve this goal.

The Dissemination Role of GREEN

GREEN acts as a network for communication and as a clearinghouse of information on water quality and environmental education. GREEN and affiliated programs have developed many educational materials and curriculum guides that contain innovative activities and documented successes from individual programs. Many of these materials are available in both English and Spanish.

International participants are encouraged to contact GREEN Country Coordinators or active GREEN programs near them to create local networks. GREEN supports collaborative work among educators with similar resources and water quality conditions. One example is the work of Rob O'Donoghue, an educator/organizer in South Africa, who has developed methods for creating "home-made," low-cost equipment that gives fairly reliable results. These low-tech test kits provide a model for educators in similar situations.

GREEN also disseminates vital information to interested groups through a series of training sessions. These workshops, which can involve teachers, students, administrators, and other community members, are available to any interested group and can be designed to fit the unique needs of the program. GREEN will soon be working with other groups to host international training workshops, where teachers and students can share monitoring data, develop action strategies, and design future cooperative projects. Such workshops help raise cultural awareness and environmental sensitivity.

Budgetary Considerations

The initial international workshops were funded by a variety of foundations, corporations, and the University of Michigan. After the summer of 1989, most of GREEN's funding has come from the General Motors Foundation, the U.S. Environmental Protection Agency, the U.S. Department of Education, the U.S. Department of Energy, Bullitt Foundation, United Nations Environment Programme, Bonneville Power Administration, Key bank, and the National Science Foundation. To date, the General Motors Foundation continues to sponsor the infrastructure of GREEN, while other proposals have been accepted for the development of specific programs within the Network. Other funding sources continue to be cultivated.

Some components of GREEN, such as the educational materials and teacher-training workshops, are self-sustaining through sales and consulting fees. Many individual catchment programs are funded by local businesses, foundations, or by school districts themselves.

Variety of Programs within GREEN

GREEN connects a wide variety of water monitoring programs around the world. In Europe, the network has linked older, established educational programs with other innovative ones. For example, skills and expertise of students in northern Italy, whose program is funded by their local government, complement the ingenuity of the program in Ecuador, where students

Figure 2.10. Russian students using school computers to communicate with partner catchments through the "Internet." Photo by Heike Mainhardt.

in a nationwide school network monitor for aquatic insects. In the Rio Grande catchment, U.S. and Mexican students participating in another GREEN project—Project del Rio—communicate through a bilingual computer conference. Students from Mexico were so enthusiastic about communicating with their American counterparts that they would travel to the local university computing center after school and on weekends—a rare example of motivation on the part of most students.

Current Needs and Direction

Currently, GREEN is involved in a variety of programs that will enhance catchment education projects worldwide. A pilot study of a toxics assessment program is underway on two catchments—the Rouge River and the Black River in Michigan, USA.

An air monitoring program, in which students learn to monitor ground level ozone, was initiated in 1991 with the assistance of the TERC in Boston, USA. GREEN is currently considering a proposal for the Nile River that would link 10 countries in a basin-wide interdisciplinary program. This

Figure 2.11. A water quality monitoring workshop on the Na'aman River near Haifa, Israel for Arab-Israeli and Jewish-Israeli students and teachers.

effort would involve an unprecedented education program designed to promote sharing and understanding between cultures.

Other hands-on approaches to classroom science which GREEN is exploring are ground truthing from satellite photographs (in cooperation with the Aspen Institute for Global Change, USA), soil monitoring, and brackish water monitoring.

Research

GREEN's programs offer a rich resource for research in environmental education; telecommunications within the classroom; cross-cultural communication among students; changes in teaching practices; collaboration between schools and communities; and school support of interdisciplinary educational projects. Student action-taking is another promising area for research that has received little attention in an educational context. The NSF-GREEN Teacher Enhancement Program is an ideal backdrop for such research efforts. Graduate students are working in the program to examine some of the aforementioned issues. GREEN will strive to find further resources for research into the effects of environmental monitoring programs on both the environment and the educational process.

This manual is another way for GREEN to promote more activities and research into alternative monitoring methods. Many nations within the network do not have the resources or access to the test kits available in the United States, Europe, and Australia. However, the components in these field kits can be duplicated with chemistry lab equipment, or replaced by other methods, such as biological monitoring (benthic macroinvertebrates, bioassay, etc.). GREEN will continue to investigate issues of cost, accuracy, and skill level for a wide range of monitoring techniques.

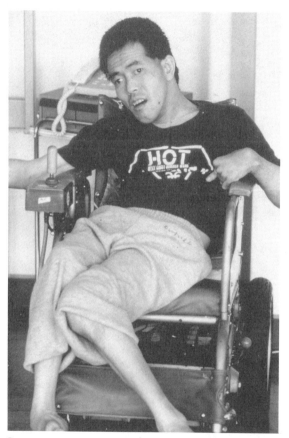

Figure 2.12. A patient with congenital Minamata disease, Minamata, Japan. The disease is caused by eating organisms contaminated with mercury.

Regional Development of GREEN

The value of a network is its relevance to local participants. Over the long term, GREEN plans to establish regional offices throughout the world to strengthen the international network. The experience of other international organizations shows that developing a decentralized regional infrastructure is the most productive and culturally sensitive way to achieve GREEN's goals.

Each office will identify regional needs and resources. It will set the priorities of the region and commit to utilizing local expertise and resources to help resolve local water quality issues.

Country coordinators will provide the backbone for this decentralized network. These coordinators are chosen to assist in the dissemination of the GREEN Newsletter, and more importantly, to determine the basic needs and priorities for their countries, and to draw upon local resources to enhance their nation's programs. They may also initiate local GREEN workshops for teachers, further disseminating and strengthening their programs.

A worldwide meeting of GREEN's country coordinators was held in 1995, to be followed by meetings on each continent. Within the U.S., GREEN is already moving toward a more decentralized structure with the opening of the Northwest Regional Office in 1994.

Figure 2.13. An Egyptian woman and child watering livestock and collecting drinking water from backwaters of the Nile River.

Conclusion

The concept of student environmental monitoring is an exciting one. The prospect of students, teachers, researchers and other professionals communicating about their local environmental concerns nationally and internationally is significant. It is GREEN's vision to reach this potential to empower students to become active learners and problem-solvers through a successful networking system incorporated into their educational process.

The broad response that GREEN has received makes it clear that educational systems around the world are ready to incorporate real-world topics into their classrooms and to encourage their students to get involved actively in learning. On a deeper level, GREEN participants appear to be ready to dissolve cultural boundaries and to open up a greater sense of understanding, cooperation, and respect between nations, especially concerning environmental issues. GREEN hopes that this will help create a more beneficial, cooperative link between cultures as the world faces serious environmental concerns today and in the future.

For all of the above to occur, low-cost monitoring techniques must be made available so that people in all regions of the world are able to monitor their catchments. This approach is also in keeping with the spirit of appropriate technology. Low-cost equipment is relatively easy to use, can be accurate, furnishes valuable information upon which to base management decisions, and provides a set of observation skills that a learner will carry for life. It is with this ideology in mind that we have prepared this manual.

Land, Rivers, People

Introduction

This chapter will discuss how the relationship among humans, their land uses, and water affects the health of the entire natural system.

Water is an essential, yet finite, substance for all human, animal and plant life. Although water covers more than 70 percent of the Earth's surface, 97 percent of this is salt water in the seas and oceans. Approximately 80 percent of Earth's freshwater is frozen in ice caps and glaciers and therefore technically unusable by Earth's biota. The usable freshwater that remains is unevenly distributed across the surface of the planet. For example, only fifteen of the world's largest rivers carry up to one-third of the total global surface water.

Water is an extremely important resource for humans. We admire it for its beauty and we need it for agricultural, municipal and industrial uses, drinking, transportation, recreation, and just about every other human activity. The problem with our usage of water is that with almost every use we contaminate or pollute it. As populations multiply and nations become more industrialized, water is being used more heavily than ever. This puts an unbearable strain on water's natural ability to cleanse itself of pollutants. The question of water quality is perhaps most acute in the developing world, where it is currently estimated that over one billion people are unable to obtain acceptable drinking water and 75 percent of the population lacks adequate sanitary facilities. Both water quality and water quantity have become critical issues, affecting all life.

The state of a local water body reflects the health of the surrounding environment. A healthy stream will reflect a healthy environment and, conversely, a polluted stream will reflect unhealthy, unwise land uses in the surrounding environment.

Rivers

Rivers have been a feature of the Earth for millions of years. Over large spans of time they have helped to shape the Earth's surface into the surrounding environment that we recognize today. Just imagine the age of the Colorado River in the United States, which has cut a channel almost 1.6 kilometers deep called the Grand Canyon. Over the years, rivers have provided habitats for the evolution of organisms that live underwater. Many orders of aquatic insects are found in rivers around the world, and many are similar as to both genus and species, whether they come from a river in Africa, the Middle East, Europe, Asia, Australia, Latin America, or North America.

Figure 3.1. A woman from Swaziland, Africa, carrying a bucket of water from a local river to her home.

Rivers were also the sites of most early human civilizations, since rivers provided freshwater, food, and the opportunity for expanding populations to move upstream. Floodplains provided rich areas to grow crops, and rivers themselves were used for transportation and to carry out domestic chores. Rivers continue to play an important role in many of the world's cultures. People began to settle and cultivate the fertile Nile Delta over 7,000 years ago, and today 95 percent of the population of Egypt live in the valley of the Nile River. In India, people consider the Ganges River holy and bathe in the river to wash away sin. The native peoples of North America traditionally celebrated the return of salmon every year to the rivers where the fish were hatched.

The River Reflects Land Uses

The total amount of freshwater on Earth is fixed. The hydrologic, or water cycle, circulates the supply of water from land and water bodies to the atmosphere and back again. Similarly, we are, in fact, continuously reusing the same water. Some stages in the hydrologic cycle, such as the intake and respiration of plants, or leaching through wetlands, help filter impurities from the water. However, increasing human use is taxing the water cycles' capacity to cleanse itself.

Land and water are inextricably connected. Almost everything you do, whether at work, at home, at school or at play, will have some impact on your catchment area. If you live upstream in a catchment area, the use you and your community makes of water resources affects your neighbors downstream. Soil erosion can lead to losses of plant productivity and/or soil stability downstream. Changes in vegetation and soils can alter streamflow, sediment flow, and water quality, affecting the health and welfare of people living downstream. As much as 70 percent of nitrates and phosphates found in downstream waters originate in the upper reaches. In many countries, such as Hungary and Egypt, more than 90 percent of water supplies originate from outside the countries' borders. In these instances, downstream nations are extremely dependent upon the actions of their neighbors.

One-quarter of the earth's land surface is mountainous and is inhabited by 10 percent of the world's population. Most of this area has a moderate amount of moisture, forest or shrub cover with little arable soil, and a low human population density. Many of these mountain regions form the headwater areas of major rivers that directly impact large populations of people living downstream. Natural and human-caused erosion in the densely populated mountains of Nepal contributes 250 million cubic meters of silt to the Gangetic Plain each year. According to observers, the beds of rivers in the Terai Plain of southern Nepal are rising 15-30 cm annually, resulting in

Figure 3.2. People enter the upper Ganges River for domestic or religious purposes.

Figure 3.3. A Nepali village recently completed a project to bring fresh mountain water to small villages as a substitute for river water.

flooding and changing river courses. Since 1850, the Kosi River, also in Nepal, has shifted its course 115 kilometers westward, leaving 15,000 square kilometers of once fertile land buried under eroded soil and rocks.

Degradation of catchment areas has already taken place over much of the Earth's surface. At one time, over 75 percent of Ethiopia was covered with forests, which moderated soil loss. Recent surveys indicate only about 3 percent remains forested. This deforestation increases soil loss and water runoff and means that groundwater aquifers are not being recharged to the same extent. Researchers estimate that only one percent of Ethiopia's water is percolating down to their aquifers. The damage during the severe 1994 flooding of the Mississippi River in the United States was exacerbated by the dams and surface impoundments built for irrigation purposes, settlement in floodplains, and some agricultural practices.

An International Example: The Nile River

The interconnectedness of land and rivers is even more critical for international rivers, that is, those that flow through more than one country. Throughout history, the Nile River has had a dominant influence on the economic, cultural, social, and political life in the ten African nations which share the river basin: Kenya, Tanzania, Burundi, Rwanda, Zaire, Uganda, Central African Republic, Ethiopia, Sudan, and Egypt. The Nile provides water for irrigation, industry, hydropower, fishing, transportation, recreation, household uses, and various municipal needs. Traditionally, the availability of water has meant the difference between prosperity and famine in the Nile Basin. The Nile Basin is experiencing a population increase which is among the highest in the world, with annual growth rates between 3 and 4 percent per year. Due to the high population increase and overextended natural resources, famine and hunger-related stress have caused hundreds of thousands of deaths during recent years.

Land is classified in the Blue Nile Basin as 37 percent agricultural, 52 percent grazing, 3 percent forest, and 5 percent bamboo forest. Land degradation and soil erosion are widespread and severe in the highlands, with an estimated 1.9 billion tons of fertile volcanic soil eroded annually and carried to the sea by highland rivers. The main causes are removal of vegetation for crops, farming on marginal lands, increased livestock pressures, shifting cultivation, and uncontrolled burning. Land degradation and soil erosion have affected water flows and storage, and increased water pollution for all downstream nations.

Environmental water problems are causing international political tensions, which are likely to increase as water in the Nile system decreases in quantity and quality. Four critical international issues confront the stability of the Nile Basin: Kenya's initiative to pump water from Lake Victoria to the

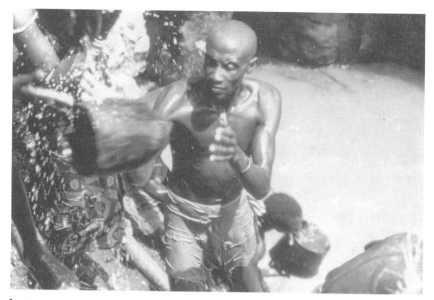

Figure 3.4. Due to expanded deforestation in Ethiopia, bore holes are used to acquire water for cattle and domestic chores.

fertile lands in northern Kenya, meaning less water for downriver communities; Ethiopia's proposal to pump water from Lake Tana to a reservoir in the cooler northern region to reduce evaporation; Egypt's agreement with Israel for the El-Salam Canal Siphon under the Suez Canal as part of the North Sinai Agricultural Development Project, resulting in reduced water for delta crops; and the Aswan High Dam, built by Russian funds, to provide storage of waters in Lake Nassar to be released to extend agricultural land by 2.5 million acres, and the number of crops per year from one to four. The warm water collected behind the dam has lead to high levels of bilharzia (human blood fluke) disease and the displacement of over 100,000 Nubian people. Each of these projects may appear to benefit the individual countries, but each has serious consequences for the basin as a whole. These and other problems make the management of the Nile Basin by all affected countries a political and environmental issue of international scope.

Catchment Area

The origins of a river are called its headwaters, and the end which drains into a pond, lake or ocean is the mouth. Rivers may begin from underground water coming to the surface (springs), from rainfall or snowmelt, as drainage from a wetland, as meltwater from a glacier, or as an outlet of a lake or pond. Depending upon the physical nature of the land, a

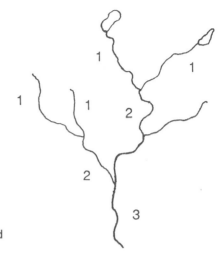

Figure 3.5. A diagram of first, second and third-order streams.

stream may be crowded with vegetation and trees along its bank. In mountainous areas, streams may only tumble over and around rocks with very little vegetation. Intermittent streams actually dry up during periods of low rainfall.

As streams increase in flow and join with other streams, a branching network is established. This network from headwater streams to the river mouth is called a river system. A numbering method has been devised to describe the relative position of a stream within a river system. The beginning of a stream is described as a first-order stream. Two first-order streams join to form a second-order stream, and two second-order streams become a third-order stream, and so on (see Figure 3.5). The Mississippi River in the United States is considered a twelfth-order river near its mouth.

The land area that drains rain and snowmelt to a river is called a catchment area. Within a catachment, each tributary is part of a smaller system. Every part of the Earth lies within a catchment area.

Water Pollutants

Many catchment areas have been altered as a result of human needs for water, food, recreation, transportation, manufactured goods, and other amenities. These growing demands have led to pollution of streams and rivers, and to unwise land uses that further degrade water quality.

Pollution sources are divided into two groups, depending on how the pollution enters a body of water. Point source pollution is waste that comes from a specific location. Factories and wastewater treatment plants may have discharge pipes that lead directly to a waterway. These are considered point sources because they are easily identified as coming from one site.

Figure 3.6. Rivers in many parts of Thailand are lined with homes. Many of the waterways receive untreated wastes and are a health hazard.

Nonpoint source pollution does not come from a specific location. Instead, it results from the runoff of water (rainfall, snowmelt, etc.) over land. As this water passes over the ground, it picks up pollutants and carries them into local streams and rivers. Nonpoint source pollution can also result from airborne pollutants that are deposited in waterways.

Nonpoint sources can be urban or rural. Nonpoint source pollution in rural areas usually results from poor agricultural or forestry practices. Urban nonpoint source pollution is caused by the runoff from city and suburban areas.

While pollution from both point and nonpoint sources can be quite harmful, our ability to control the two types of sources is very different. Pollution from point sources is much easier to limit because its origin can be readily identified. The responsible company or factory can be contacted and asked to reduce its discharge. In some countries, stronger laws can be passed to limit the discharge from point sources.

Nonpoint sources are more difficult to control. They are harder to identify, and may be the result of land use practices across an entire catchment area. Examples include the contamination of streams and rivers by fertilizers and pesticides used on lawns in suburban areas and on crops and fields in agricultural areas. Both types of pollution are challenging to control because there may be hundreds or thousands of widely scattered sources within a catchment. Source reduction of nonpoint source pollution may require that those living in the catchment area change their behavioral

Figure 3.7. In this industrialized sector of Taiwan, a major source of water pollution is air contaminants.

patterns and lifestyles. This is more difficult than trying to change the action of an identifiable source polluter. Also, nonpoint sources are not regulated by law as stringently as point sources. A key element in controlling nonpoint source pollution is for individuals to take responsibility for their own actions.

Water Pollution Categories

There are four major categories of pollutants, and each has different direct or indirect effects on the environment.

Organic Pollution

Organic pollutants come from the decomposition of living organisms, either plants or animals, and their by-products. Grass clippings, leaves, human sewage, and pet wastes are examples of organic pollution. Organic pollutants use oxygen from the aquatic ecosystem in the decomposition process. When there is an excess of organic pollutants, they can deplete the oxygen in the system, making the river an anaerobic system.

Figure 3.8. In Australia, Taiwan and other regions of the world, river bottoms are sometimes mined for sand and gravel during the dry seasons.

Inorganic Pollution

Inorganic pollution consists of suspended and dissolved solids, such as silts, salts, and other minerals carried into streams from streets or exposed soil. Common sources of inorganic pollution include the runoff from roads that have been salted or sanded, construction sites, and plowed croplands. This type of pollution can cover up the eggs of spawning fish and suffocate other plant and animal life, causing a decline in the diversity of the ecosystem.

Toxic Pollution

Toxic pollutants are heavy metals (such as cadmium, mercury, chromium, iron and lead) and organic compounds (PCBs and DDT) that are lethal to organisms or interfere with their normal biological processes at certain concentrations. Toxic pollutants are often produced as by-products of industrial processes. Another source is household products, such as bleach, drain cleaners, and pesticides. Street runoff and airborne contaminants can carry toxins into waterways. Standard farming practices contribute herbicides and insecticides to surface waters.

Thermal Pollution

Thermal pollution is "waste heat." It is often a consequence of using water to cool industrial or power generation processes, and then returning it at a much higher temperature into local waterways. A nonpoint source of thermal pollution comes from urban runoff. Smaller streams are more vulnerable to thermal pollution than large rivers. Thermal pollution increases water temperatures, speeds up the life cycles of animals, and affects the food sources of migratory birds. It also affects the dissolved oxygen level of the rivers, since less oxygen dissolves in warm water.

Land Use Practices

How urban, agricultural and forest lands are managed and developed influences local water quality in many ways.

Urbanization

Urban areas are places where humans live together in relatively dense conditions and have created enormous changes in the natural environment. Most of the natural vegetation and rain-absorbing soils are replaced by impermeable surfaces such as roads, parking lots, sidewalks, and rooftops. Streams are often encased in concrete pipes and transformed into underground storm sewers. Natural watercourses are altered as floodplain areas are converted to industrial, commercial, and residential uses.

As natural areas become covered by impermeable surfaces, less rainfall is absorbed by the ground; runoff increases and can create flood conditions. Storm runoff is often warmed as it passes over pavement and other surfaces, leading to warmer water temperatures in rivers and streams. Removing trees from stream banks also causes water temperature to rise, as does the increased turbidity caused by urban runoff.

Depending on the current, the silt carried by urban runoff can scour the stream bottom in some places and smother it in other places. It may be carrying heavy metals and other pollutants with it. Other toxins, such as used paints, solvents, cleaners, and oils, are dumped by people into sinks or storm drains and eventually flow into local waterways. All of these pollutants have severe effects on aquatic life.

As urban areas grow, the amount of pollutants washed into storm sewers and carried into local streams also grows. In urban areas with separate sewer systems, untreated runoff from precipitation travels through storm sewers directly into waterways. Combined sewer overflow systems (CSOs) (see Figure 3.10) operate differently. In CSOs, sewage and storm runoff are carried by the same pipes and treated together at wastewater treatment plants. During periods of heavy rains, pipes may become overloaded and discharge large volumes of untreated sewage mixed with street runoff directly into waterways, resulting in severe water quality problems.

Figure 3.9. The mouth of the Cuyahoga River is an outlet to Lake Erie and is used to transport needed resources. This sector of the river caught fire in 1969.

To remedy the situation, some treatment plants are building retention tanks or basins to store untreated wastewater during heavy flows, to be gradually released for treatment later.

Malfunctioning septic systems along rivers, illegal sanitary connections, lack of sanitary treatment or improperly treated wastes, and the influx of raw sewage from CSOs contribute bacteria and nutrients into rivers, leading to health problems and cultural eutrophication downstream.

Unsound Agricultural Practices

On a global basis, agriculture currently accounts for almost 73 percent of total freshwater usage. About 18 percent of the Earth's cropland is now irrigated, producing about 30 percent of the world's food. However, as the use of irrigation increases, environmental costs also increase. Increasing water withdrawals for agricultural use is depleting aquifers, and is reducing or eliminating crop production in some fertile regions through salinization, alkalization, and water logging. Our society depends on the farmers who feed us all. However, poor agricultural practices can harm water quality as well as threaten the fertility of soils.

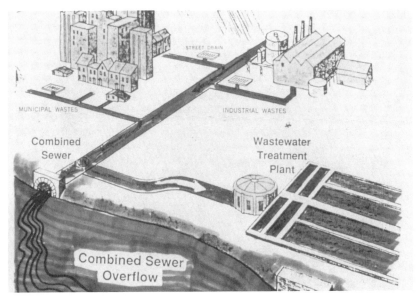

Figure 3.10. This diagram illustrates a Combined Sewer Overflow (CSO). During heavy rains, discharges to the river occur to prevent flooding of the waste water treatment facility.

Agricultural pollution of surface-water bodies takes three general forms: (1) sedimentation from soil erosion; (2) eutrophication, or nutrient enrichment by nitrogen and phosphorus; and (3) contamination from toxic chemicals such as herbicides and insecticides, or from disease organisms.

A variety of agricultural practices can cause rural nonpoint pollution: leaving topsoil exposed where it can be eroded; overgrazing pastures; compacting soils; leaving streambanks unprotected; overusing insecticides, herbicides, and fertilizers; allowing liquid wastes to escape from feedlots; and removing windbreaks. Many of these practices allow fertile topsoil to be carried into local streams, adding silt, toxic pollutants, and nutrients to the water, and disrupting the ecology of aquatic communities. Not only does soil erosion cause drastic changes in the quality of water and the aquatic community, it may also have serious nutritional effects as soil fertility, and therefore crop production, decreases. Furthermore, siltation, which is mostly caused by soil erosion, can have drastic economic effects as well. For example, Argentina spends $10 million per year to dredge silt from the estuary of the River Plate.

Salinization, the build-up of salts in soil and water, is another serious problem, especially in irrigated areas. All water unless it is distilled or deionized, contains salts. In arid regions, natural salts concentrate in the surface of the soil as irrigation water evaporates or is taken up by plants. Salt levels in the soil gradually become too high to support plant growth, and

Figure 3.11. The waters of the Danube originate in southern Germany, and pass through major cities, such as Vienna, Belgrade, Budapest, and Bucharest. What city is this?

the land becomes unusable for farming. Not surprisingly, the runoff from these lands can be very saline.

Excessive soil salinity can also be caused by irrigation water seeping down through the soil and raising the height of the water table. The salts in underground rocks are dissolved and carried upward as the water table rises, eventually reaching root systems and killing plants intolerant of saline conditions. This leads to wind and water erosion, since dead plants are no longer capable of binding the soil. Timber clearing in the upper catchment area can also cause dramatic rises in the water table, with the same results.

Dry land salinity may be caused by reduced transpiration from an introduced vegetation type. For example, deep rooted trees have been replaced in the Murray-Darling river system in Australia by seasonal crops or grasses that do not pump the water into the atmosphere as efficiently.

Many farmers are now using alternatives to these destructive farming practices. Windbreaks and no-till farming help prevent excess runoff and soil erosion (although no-till methods may require the increased use of pesticides). Leaving forested buffers between fields and streams helps to reduce runoff dramatically. Many farmers are converting to organic methods, reducing their use of insecticides, herbicides and inorganic fertilizers, and replacing them with natural fertilizers such as manure, and biological methods of pest control. These alternatives are finding more appeal with farmers as they learn the benefits of protecting soils and streams.

Unwise Forestry Practices

The harvesting of trees can influence water quality. Road construction, clear-cutting steep slopes, and dragging the trunks of the trees into rivers to float to sawmills have all had serious effects on rivers. Globally, catchment area forests are being widely devastated. Deforestation is occurring rapidly in many countries as forests are cleared for agriculture, cattle grazing, fuel, fodder for animals, and foreign trade.

One of the worst consequences of poor forestry practices is the erosion that follows the ground disturbance and removal of trees associated with logging. This has destroyed benthic communities and ruined the spawning beds of certain fish species. Soil erosion from unsound forestry practices is causing the suffocation and death of corals in the ocean surrounding the Philippines and in the Great Barrier Reef of Australia. The clearing of forests may also drastically shorten the economic life of dams and reservoirs further downstream by increasing the sediment backed up behind an impoundment structure. Deforestation in Northern Luzon, Philippines, has reduced the life of dams in the area from 60 to 32 years.

Fortunately, in many countries, foresters are becoming more conscious of environmental effects, and guidelines have been established to protect ecosystems from logging damage. As this understanding increases, the knowledge and lessons can be passed on to citizens in other countries.

Waterway Alterations

Dams and other modifications provide benefits to humans, but they can affect stream ecology dramatically.

Dams and Impoundments

Dams are constructed to provide community needs for cheap hydro-electric power, flood control, irrigation, drinking water and recreation.

Because there are few remaining free-flowing rivers, the ecological impacts caused by dams and their impoundments should be understood. A dam transforms a flowing stream or river into a lake-like environment. The current slows and the water warms, leading to reduced oxygen levels and a more favorable environment for the breeding of pathogens.

Figure 3.12. Rivers in many regions of the world, such as the Niger River in Nigeria, are used to transport logs to cities.

Behind the dam, rooted aquatic plants flourish along the shoreline, and some species of dragonflies and damselflies, water striders and giant water bugs, and chironomid larva may become more abundant. However, due to the warmer water, some of the organisms that demand higher oxygen levels, such as mayflies, stoneflies and caddisflies, may decrease.

As water is periodically released from the dam, erosion may be accelerated, and the diversity of aquatic life decreased.

Channelization

Another method of controlling river flow is channelization, a process in which rivers are dug out, straightened, and sometimes lined with concrete. Trees and shrubs along the river are often removed. Rivers are channelized for flood protection, farmland drainage, and sometimes for navigation.

The ecological impacts of channelization can be significant. Straightening of the river causes the speed of the current to increase, leading to greater flooding downstream. The erosion of the straightened river banks increases turbidity. In low water, the channel becomes choked with mud, creating an unstable substrate for benthic life. Water temperatures rise, food sources are eliminated, and the river becomes a harsh environment. Aquatic diversity drops sharply.

Figure 3.13. Channelization is often used to move water through urban areas to reduce flooding or to deepen the river for shipping.

Channelized rivers in urban areas often become conduits for waste, and are considered community health hazards where access is restricted by tall chain-link fences. Natural, tree-lined, winding streams are converted into straight, exposed ditches—unattractive and hostile environments for people and aquatic life alike.

Wetland Drainage

Wetlands are land areas, such as marshes, swamps, and bogs, that are perpetually wet, yet shallow enough to permit standing vegetation. Wetlands are some of the least understood and most threatened aquatic habitats. In the United States, wetlands have been widely drained and destroyed in the process of converting land to other uses. River channelization, the expansion of agriculture, and urbanization have all contributed to the loss of wetlands.

Wetlands are valuable resources for many reasons. They provide important habitat for waterfowl, fish, and other wildlife. They help control floods by slowing and retaining runoff. They also function as groundwater recharge sites by allowing water to percolate through the moist soil. Wetland vegetation traps the sediment in runoff and absorbs nutrients. Because of this ability to absorb nutrients, artificial wetlands have been constructed by some communities as an alternative form of sewage treatment.

Because wetlands are so important for wildlife, their loss has caused some species to decline to the point of extinction. Fortunately, recognition of the value of wetlands is growing and many are now legally protected. Many previously drained wetlands are being restored in efforts to recover the benefits of these invaluable ecosystems.

Land, Rivers, People

Water sometimes seems to be an infinite resource. However, the use and abuse of water resources by a burgeoning human population and an ever-expanding industrialized society, have begun to seriously affect the very nature of water and the water cycle itself. Increasing global urbanization is placing a great stress on both surface and underground water resources. Degraded water quality in the developing world is a severe problem that will have to be tackled before any kind of real development, whether economic or qualitative, can be accomplished.

Currently, 2.4 billion people in the world live in urban areas. By the year 2025, the world's urban population is expected to reach 5 billion, putting additional strains on both water quality and quantity and increasing stress on an already scanty infrastructure. The management of water and catchment areas is a global issue. There are more than 200 international river basins—57 in Africa, 48 in Europe, 40 in Asia, 36 in South America and

Figure 3.14. Wetlands serve as habitat for wildlife and fish, help control floods, absorb nutrients, function as ground water discharge, and trap sediments.

33 in North and Central America. The lives and livelihoods of half of the world are directly dependent on the way in which catchment ecosystems within these international river basins are managed.

Throughout history, joint use of catchment basins has always depended on cooperation among riparian states. Failure to reconcile the competing interests of upstream and downstream users has generated considerable political friction in many parts of the world. This is nowhere more apparent than in the arid Middle East, where nothing is more important than water. In the states on the Arabian Peninsula south of the borders of Iraq and Jordan (including the Kingdom of Saudi Arabia, Kuwait, the United Arab Emirates, Bahrain, Qatar, Oman, and Yemen), there is not a single stream that flows the year around. The 23.3 million people living in this area are completely dependent upon underground aquifers. Underground aquifers do not exist in isolation from the hydrological cycle. They must be recharged by precipitation elsewhere in the catchment area. If water is irresponsibly managed in the upper reaches of the catchment, the lives of millions of people in this area will be negatively affected. The citizens of the Arabian Peninsula must practice effective water management themselves and they must work in cooperation with nations in the headwater regions, such as Turkey and Syria, in order to guarantee their future survival.

Most citizens in industrialized countries have access to safe drinking water. But this does not mean that they no longer have to worry about monitoring their water quality. Point source pollution has been decreasing in many countries, but nonpoint sources continue to cause many problems, especially with groundwater supplies. The connection between groundwater and surface water is sometimes hard to visualize, but the two are related. The water that flows over land and into the rivers also percolates down through the soil to recharge deep groundwater aquifers. Land use practices that are polluting rivers and streams have the potential to pollute these groundwater sources. The industrialized countries of Europe and North America face increasing contamination of groundwater reserves. In the United States, up to two percent of the deep aquifers may be unsafe to use for drinking water. Pollution from nonpoint sources such as fertilizers and pesticide residues in farm runoff, de-icing salts in runoff from city streets and highways, leaking underground sewer lines, underground and surface mines, and inadequate disposal of chemical and other hazardous wastes are mostly to blame. In addition, many arid regions are mining their deep aquifers. As aquifers are generally recharged at a very slow rate, continuous groundwater mining can lead to land subsidence and an immense water shortage for future generations. Due to aquifer mining, a land area of 225 square kilometers in Mexico City has dropped up to 9 meters from 1891 to 1978.

Figure 3.15 and 3.16 (above and opposite). A potential health problem in this Mexican river is the release of untreated sewage from homes (above) which is used for domestic chores 100 meters down river (opposite).

Why Monitor?

Many of the problems associated with unwise land use and water pollution have been laid out above. A growing awareness is arising of the need to monitor local rivers in an effort to protect the environment and improve the quality of life for people around the world. Monitoring will help people to recognize how local land and water use affects the quality and quantity of their water resources and the resources of neighbors downstream. Once these issues are recognized, people can start to work towards local solutions.

Everyone has a right to know and understand the quality of their environment. Monitoring will help increase first-hand knowledge and experience, and prompt us to protect and preserve our water and the environment

Figure 3.17. Due to the actions taken by the United Nations and the Mediterranean Member Nations, the water quality is improving and fish production has increased.

in which we live. Continual monitoring of the same section of a river reach or catchment will also help people notice changes in their catchment area and allow them to identify potential problems. Involving students and citizens helps to empower the community, setting the stage for action taking to improve local water quality (see Chapter 9). Humans want to feel needed. By involving the public, people begin to feel that they can and will make a difference.

A good example of community involvement is the water quality program, "Waterwatch," begun by one high school monitoring the headwaters of the Mary River in Queensland, Australia, in 1989. Presently, most schools, multiple community groups, and governmental bodies are involved along the entire 305-kilometer stretch of the Mary River. To help stem stream bank erosion, local high school students and community groups drew up a plan to help stabilize the banks by removing non-native vegetation and replanting the bank with native species of trees. This project proved to be very successful.

Popular participation is an essential ingredient to successful action. A framework in which upstream communities respect the natural rights of downstream communities to an unpolluted water supply can be provided by a regional action plan. Elements of an action plan may be to: increase sustainable development, instill environmental quality to sustain the health and well being of the surrounding environment, develop a basin-wide network

Figure 3.18. The land around this well has been heavily grazed in Niger, Africa. The result is a dropping water level which threatens the village water source.

Figure 3.19. A GREEN workshop on water quality monitoring in Minamata, Japan. This is the site where Minamata disease was first studied.

Figure 3.20. The study of macroinvertebrates is a low-cost water quality monitoring technique.

for the monitoring and assessment of resources, and provide national planners with information necessary to make environmentally sound decisions. GREEN was created, in part, to help countries develop and set up action plans and international networks in an effort to improve the environment.

Need for Low-Cost Monitoring Techniques

Until recently, water quality monitoring has been done in both developed and developing countries by technically trained professionals using commercial kits. However, the data and reports of technical experts are often incomprehensible to the average public. In addition, commercial kits are often too expensive and the materials difficult or impossible to obtain for groups in many areas of the world. Some commercial kits are fairly easy to use but may still require some prior knowledge or expertise to interpret the results. In contrast, low-cost water quality monitoring techniques are available to nearly everyone. But perhaps one of the most important reasons for monitoring rivers and streams with low-cost techniques is that they will help develop our natural senses of sight, sound, and smell. Low-cost or no-cost techniques help participants develop a holistic ecological sense of their catchment area and their surrounding environment.

GREEN foresaw a need to help develop and test low-cost monitoring techniques along with other programs, particularly in poorer areas. We see these techniques as the most appropriate technology for some groups and the best suited for some situations. Implementation of low-cost monitoring is an excellent way to begin development of educational programs on a catchment-wide basis. Participation by more children and people of all ages in low-cost water quality monitoring may set the stage for appropriate actions to help improve local environments and increase the global quality of life.

Rivers of the World

Introduction

The Amazon, the Zaire, the Nile, the Yangtze, the Ganges, the Murray-Darling, the Volga, the Danube, and the Mississippi are some of the great rivers of the world. Each river is in some ways a unique reflection of its catchment. The Amazon gathers tributaries from an area the size of Australia and at its mouth empties 20 percent of the world's freshwater into the Atlantic Ocean. The Zaire is the only major river in the world that traverses both sides of the equator, and unlike many rivers, has large waterfalls near its mouth as it spills off the plateau of Africa on its way to the Atlantic Ocean. The Ganges begins its journey to the Indian Ocean as glacial melt-water high in the Himalaya mountains and forms the largest river delta in

Figure 4.1. The cataracts on the Nile River are determined and shaped by elevation and human structures.

the world at its mouth in Bangladesh. The processes that have formed these great rivers never rest, and the rivers themselves are constantly changing, although people, locked into a tiny sliver of geologic time, might view rivers as unchanging. Perhaps that is why people are drawn to rivers—because they have this dual nature; the apparent constancy of flowing water on the one hand, with their ever-changing, responsiveness on the other hand.

Rivers are dynamic, often traversing regions of varying topography, geology, climate, and ecosystems. Each river reflects a combination of physical, biological and chemical processes. Although each catchment and its river is unique, the interaction between land and water follows physical processes that allow one to understand the forces that create river systems, and to predict how changes to the river channel and the surrounding landscape impact the flow of water, its sediment load, and the river channel ecosystem.

Shaping River Systems—Climate and Geology

Two principal factors directly influence river systems. Climate prescribes both the amount of precipitation, the source of water that falls on land; and the air temperature, which determines the physical state of that water (snow, rain, sleet, or hail). Geology defines topography—mountains, plateaus, lowlands, and other landform features—which determine where and how water flows over land to begin carving channels. Geology also defines lithology, or the type of rocks that form soils in an area, in conjunction with climate and vegetation. Climate, geology, and soil character in turn shape land use and the types of vegetation found in an area, ranging from tropical forests to grasslands to arid deserts to cold arctic tundra. The effects of climate and geology interact to produce two important characteristics of all river systems: discharge, or the amount of water flowing in the channel; and sediment load, which is the solid and dissolved material carried downstream by the water (Dunne and Leopold 1978). Together, all these factors shape the character of a river and the way that water flows throughout its course.

The amount of water available for riverflow is governed by the hydrological cycle or water balance, and can be viewed mathematically: *precipitation = evaporation + riverflow + storage*. The size of rivers and their catchment areas, which include all surface and subsurface water that drains to a single point, show an unequal global distribution, because of variable distribution of the three processes regulating the hydrologic cycle—precipitation, storage as groundwater, and evaporation.

Figure 4.2. Siltation can reduce light penetration, thereby decreasing the oxygen level and increasing water temperature.

Climate and Rivers

Climate influences the amount and type of precipitation—rain, snow and condensation—delivered to land masses. The quantity and distribution of precipitation to a catchment is affected by regional climate as well as by the proximity and location of oceans and currents, latitude, landmass size, and elevation (Hartmann 1994). Areas with greater precipitation generally have river systems of greater discharge, and thus rivers with the greatest runoff generally occur in tropical and subtropical areas (Allan 1995).

The hydraulic characteristics (size and flow) of rivers are largely influenced by climate, geology and landmass characteristics, such as slope and elevation, precipitation, and vegetation. For example, the Yangtze River drains 1,942,500 km^2 of the Tibetan Plateau (Leopold 1994). It is fed by monsoon precipitation, and has a discharge of 33,111 m^3/sec. To visualize this kind of discharge, think of 33,111 boxes, each a meter on a side, moving past you every second! Thought of as runoff, this would be .017 m^3/sec for each square km. The adjoining basin to the north, the Yellow Yellow River, drains 673,400 km^2 and has a discharge of 49 m^3/sec, or only 0.003 m^3/sec for each square kilometer. It drains a cold, dry steppe, and flows through part of the great loess deposits of that arid region.

Water delivered to a landmass may be stored within plant cells and the soil, on ground surfaces and in depressions, as groundwater, or added to streams and lakes. The distribution of water in these storage areas varies

Figure 4.3. Removing the vegetation from the surface of rivers can increase the flow of water and reduce evapo-transpiration.

widely, depending on geology, soil types, plant ecosystems, and air temperature. The rate at which these storage units function to retain or release water over time depends upon climate. Outputs include evaporation which is regulated by temperature and relative humidity, and plant transpiration which is dependent on plant species, density, structure and distribution. Human activity can affect this system. Removal of vegetation (Fig. 4.3), for example, reduces the rate of evaporation, resulting in increased riverflow (e.g., Swank et al. 1988). On a large scale, however, massive deforestation changes precipitation patterns and often causes decreased river discharges.

Geology and Rivers

Geology has a profound influence on rivers and their catchments. The movement and erosion of land over time creates physical and chemical boundaries with which water interacts. Natural features like mountains, valleys, and plains directly affect how, and in what direction, water flows. Elevation and river slope control the power of water, as it drains toward a lower point, influencing catchment size and form. The velocity of water and its capacity to erode are dependent on the existing terrain. In steep terrain, the velocity of water and its capacity to erode are greater than in flatter areas. Patterns of precipitation, presence of vegetation, and soil character also define the capacity of water to erode the land and to create river channels and carry sediment. Rainwater runoff and snowmelt in some regions carry bound-up minerals from land into the water column which influences the chemistry of the river.

Soil Character and Vegetation

The interactions between climate (temperature and precipitation), geology (elevation and latitude), lithology (chemical and physical properties of mineral material), living organisms (animals, plants, and microorganisms), and time (duration over which climate and organisms have influenced the parent material) help create soil regimes. Soils continually interface between atmosphere and river channel, imparting to river communities much of their ionic chemistry (ions such as Na^+, Cl^-, Ca^{++}). Further, soil texture influences resistance to erosion as well as the type of plant community. Soils are grouped according to properties such as moisture, temperature, color, texture, and structure. Chemical and mineral properties which define soils, include levels of organic matter, clay, iron and aluminumoxides, silicate clays, salts, and pH. Each soil group has varying capacities to support vegetation (including agricultural crops), and each reacts differently to temperature change and precipitation.

Geology and vegetation together determine how rainfall or snowmelt is stored and moved into the river channel. The rate that water infiltrates the land surface is dependent on soil properties such as pore size and soil texture. Porous soils, such as sands, have higher rates of infiltration than compact/dense soils such as clays or bedrock, in which water may infiltrate slowly or not at all.

Flow to the River

Water is supplied to a river channel in four ways: channel interception, surface flow, subsurface flow, and via groundwater. The most direct route is channel interception, in which rain or snow falls directly into the river. This usually contributes an insignificant percentage of total water within the channel, but can be important in larger river systems. Surface flow is generated when there is no infiltration capacity by the surrounding land, such as when the surface is bedrock, or the soil is already saturated with water.

Water also is supplied to river channels through subsurface sources. Subsurface flow occurs when water infiltrating into a slope meets soils of low porosity; water then flows downslope. Groundwater is another source of river water and enters the river at the point where the channel cuts below the water table. Groundwater reaches the stream channel slowly over long periods of time, sustaining streamflow during rainless periods (Dunne and

Leopold 1978). Artesian waters, pumped into the channel through springs and seeps that are under pressure, are another source of water for rivers.

Differences in flow allow us to characterize types of river systems. At one extreme are rivers fed by precipitation; at the other extreme are groundwater-fed rivers. In the former, water sources include direct precipitation and surface flow; in the latter, sources are subsurface and groundwater flow. Almost all rivers lie somewhere between these extremes. Typically, groundwater moves at a rate much slower than that of subsurface and surface flows. Thus, flows in groundwater-fed streams are often stable, whereas surface and subsurface-fed streams tend to show peaks and lulls in flows. For example, two streams in Michigan, USA, generally receive equal amounts of precipitation, yet display markedly different flow rates. The River Raisin catchment, in southeastern Michigan, is overlain by till and fine lake plain clays with low porosity. Over the course of a typical year, daily discharges can vary by as much as four orders of magnitude. A large proportion of discharge is made up of stormflow contributed by surface and subsurface runoff, and peak flows—the highest level of flow after a precipitation event—are very high. The Au Sable River, in northern Michigan, drains an area predominately underlain by deep layers of sand outwash, characterized by high porosity. A much greater proportion of discharge is the result of groundwater, and the daily discharges vary little more than a factor of one.

Even in the absence of an aquifer, or area of groundwater, drainage from a soil profile can maintain streamflow over long periods in humid areas (Hewlett and Hibbert 1967, cited in Gustard 1992). Nevertheless, many catchments lack significant storage of groundwater. Where rainfall is relatively uniform throughout the year, large rates of evaporation in the summer mean that many headwater catchments have a characteristic drought season, even in humid climates. Agricultural draw-down and development of wetlands can intensify this problem by lowering the water table.

Climate and geology also affect thermal characteristics of river systems. Stream water temperature is affected by the route water reaches the stream channel, and by the ability of air or the ground to cool or warm that water. Rivers that have stable flows can differ significantly in flora and fauna due to water temperature variations.

Discharge

Rivers are responsive to changes within the catchment, seeking equilibrium among factors like rainfall and structure of the river channel. Morisawa (1985) calls this a process/response system, linked so that a change in any part of the catchment causes a response elsewhere. If some change in the catchment causes an increase in discharge, there will be a corresponding increase in stream velocity. This, in turn, leads to greater erosion (e.g., the river cuts a wider channel) that decreases the velocity over time so that erosion decreases and a new balance or equilibrium is achieved between the size of the river channel and hydrologic conditions.

Climate, geology, soils, and vegetation all help create different kinds of river flow regimes around the world. Beckinsale (1969, cited in Burt 1992) described catchments around the world that closely reflected regional patterns of precipitation, temperature and evapotranspiration. These are listed in the chart on the next page.

Each of these major groups can also be tied to the periodic nature of flow. For example, a river in the tropical rain forest of central Zaire has no low-flow periods. A taiga region in eastern Siberia has cold interiors where most rivers are frozen in winter and maximum flow peaks occur in the summer.

Sediment Load

Flow also erodes sediments from land and within the stream channel, and it is this function that changes the river structure as sediments are deposited downstream. There are two important types of sediments found in flow: dissolved load and sediment load. The dissolved load includes the cations and anions dissolved in water, and is related to catchment lithology and biological activities which release chemical substances into the river system. Although dissolved load can be a substantial percentage of total load and affects river ecology, it plays a minor role in the physical transformation of the river channel. It is the sediment load which contains particles of varying size that are moved or suspended by flow. This suspended sediment directly affects stream habitats by eroding the channel and by settling of the sediment in slower reaches of the river.

There are two main sources of sediment: the channel itself, and the catchment. Observations of water flow and catchment characteristics will show which source has the largest impact on total sediment transport. Sediment movement is a function of stream power (the ability of water to move sediment) and the rate at which sediment is supplied to the river channel. The power available to transport sediment is related to discharge and slope. Greater discharge and steep slopes both generate more power to carry sediments.

Because smaller particles tend to fall into the laminar flow layer, a region of very smooth (i.e., non-turbulent) flow around solid surfaces, more power is required to move them. Effects of laminar flow can be observed even in streams with strong currents, as this region of smooth flow allows algae to colonize rock surfaces and certain macroinvertebrates to remain attached (Fig. 4.4). With larger particles, more power is required to move larger mass.

Major Flow Regimes	Rivers and Regions
Tropical rainy climate; more precipitation than evaporation	Zaire River/Africa; Rivers in Malaysia; Amazon River/ South America; Lower Ganges River/India; Mekong River/Asia Brisbane River/Austrialia
Dry climates with runoff confined to infrequent heavy precipitation	Upper Niger River/Mali; Lower Nile/Sudan; Rio Bravo/Mexico; Diamantina River/Australia.
Warm temperate climates; temperature and evaporation vary seasonally	Rhine River/France and Germany; Danube River/Europe; Murray-Darling River/Australia; Mississippi and Ohio Rivers/ United States.
Cool temperate climates; temperature and evaporation vary seasonally	MacKenzie River/Canada; Volga River/Russia; St. Lawrence River/ United States and Canada; Yenissey River/Siberia.
High altitude; water release by snow packs and glacial ice	Headwaters of Ganges River/ India; Headwaters of Yangtze River/Tibet.

Figure 4.4. Slow current allows algae to colonize rock surfaces and macroin-vertebrates to remain attached to rocks and plant life.

Studies of river channels predict that sand is the most easily eroded sediment fraction. Silt, clay, and finer particles form what is called washload, because these particles fall between larger sediments like sand, gravel, and cobble (Fig. 4.5). The washload is less easily eroded than sand because the particles must first be "raised" from among other sediments. Coarser fractions (sand, gravel, cobbles, and boulders) are termed bed (material) load, and are transported only during high flows.

A graded river is one where a balance occurs between the physical characteristics of the channel (slope and roughness of bottom), forces of water in the channel (discharge), and transport in the channel (sediment load). A graded river does not change very quickly in physical structure. Factors influencing in-stream erosion include flow properties, bank material composition, climate, subsurface conditions (e.g. seepage forces, etc.), channel shape, biology (vegetation, animal burrows), and human-induced factors (urbanization, land drainage, reservoir development).

Figure 4.5. Fine sediments, such as silt and clay, get washed through a strong rapid reach and deposited as the current decreases. The kinds of macroinvertebrates will be different in these two zones.

Structure and Function of Rivers

A myriad of factors work to form river channels and the ecosystems they influence. Rivers respond to changes in discharge and sediment amount or type, and adjust their channel form to achieve a balance in three ways. First, rivers adjust their cross-sectional shape. As discharge increases, the channel erodes and becomes deeper or wider; as discharge decreases, the channel becomes shallower with deposited sediments. Second, river beds change in terms of roughness, which in turn affects resistance to flow. These interactions lead to adjustments in the river bedform, such as dune and sandbar formation, which increases surface roughness, slows discharge, and creates riffle-pool sequences. Finally, the channel pattern itself can be changed. Channels that transport fine particles are usually deep and narrow, whereas channels transporting coarser particles are shallow and wide.

It is the river's sediment, eroded from the catchment, transported within the channel, and deposited in downstream areas, that forms the river course and the habitats in which aquatic organisms live (Fig. 4.6). Present channel form is often a function of past flow conditions, rather than the dynamics of adjusting flow.

There are three primary channel forms: straight, meandering, and braided.

Figure 4.6. Sediments carried in a river system are deposited at the mouth of rivers as the current slows down, leading to the development of deltas.

Straight Channels

Straight channels can form when the channel transports principally suspended load, with fewer large particles that can affect stream flow and build river structures, over low slopes.

Meandering Channels

The deepest part of the stream cross-section, called the channel thalwegs, moves back and forth within the channel as slope increases. With increased slope the thalweg captures more and more of the flow until the thalweg is the only place where the water is flowing. The effect of meandering is to change the slope of the river by increasing the distance required to move sediment. To travel the same distance with the same discharge, increased meandering results in less power and less sediment transported. Thus, there is more opportunity for sediment to fall out and form channel structures such as sand bars and floodplains.

Braided Channels

Braided channels, on the other hand, are associated with high discharges and high gradients. Braiding is associated with large rates of power dissipation: Instead of one large channel, many little ones form, and power can be expended in many channels in the same amount of area. Sediment loads affect the rate at which a river takes on its course. With increasingly coarse sediment particles, channels increase in width given fixed slopes and discharges (power). Larger materials require higher velocities at the bed to move the material; without changing power, the stream must change shape (shallower and wider).

River Habitats

The processes that shape river structure and influence the nature of flow within these channels also create various habitats in which flora and fauna live.

The riffle-pool sequence forms as part of a regularly-spaced alteration between areas of erosion and deposition within the stream channel. Riffles are characterized by coarser substrates, such as gravel and cobble, and shallow water depth and higher velocity. Pools are characterized by finer substrates, greater depth and lower velocity (Fig. 4.7). Riffles are areas of particulate deposition, where a gravel bar is maintained alternately on one side and then on the other side of a channel (Dunne and Leopold 1978). Because

Figure 4.7. Pools are often created in river systems by fallen logs, resulting in reduced stream flow and the dropping of river sediments.

water must rise over riffles, mean water velocity is higher due to the restricted cross sectional area. Greater current velocity and coarser substrates often mean that riffles tend to support higher densities and diversity of benthic invertebrates, and thus are important food-producing areas for fish (Gordon et al. 1991).

Floodplains are flat areas that result from the river's inability to carry all of its sediment load during flooding events that occur generally once a year and exceed the channel's banks. Active floodplain formation occurs as adjustments to the channel cause meandering within the valley. Downcutting occurs on the outward wall, and deposition on the inner bend of the meandering stream bed. Over long periods of time, the inner bend— with slower velocity—fills with fine particulates along a flat plain. Floodplains themselves are periodically inundated with floodwaters, which often results in very high rates of primary production, because the flood supports both aquatic productivity during the flood stage, and terrestrial plant productivity during the dry stage. Human development of floodplains, particularly for agriculture, flood control, and channelization, can have tremendous negative impacts on a river's aquatic biota and the surrounding terrestrial ecosystem.

Flow and channel characteristics are also the forces creating other important stream habitats, such as depositional areas of low flow, log jams and other concentrations of large organic matter, and variably shaped stream banks and undercuts.

Understanding River Ecosystems

River systems are the product of ongoing climatic, physical, biological, and chemical interactions that occur over time. However, rivers are directly affected by human activities in the channel and indirectly by land use activities. Lotic organisms must adapt to natural discharge patterns, diverse substrates, temperature fluctuations, and to biological interactions (predation and competition). Their density and distribution may be strongly impacted by human-caused disturbances to river hydrology and habitat.

To better understand the physical nature of rivers and their catchments and to assess physical factors such as discharge, gradient, temperature, substrate, sediment load, channel structure, frequency of flooding, and bank vegetation, follow the activities noted below from Chapters 7 and 8 in this manual. All activities are from Chapter 7, unless preceded by an 8 (e.g., Activity 8.5).

CONCEPT	ACTIVITY	ACTIVITY NO.
Discharge	Measure width/depth of river	7.9
	Calculate discharge	7.4
Gradient	Calculate gradient	7.4
Channel slope	Measure slope	7.9
Temperature (water and air)	Measuring temperature	7.9, 8.5
Precipitation	Recording weather, measuring precipitation	7.9
Roughness of river bottom	Substrate survey	7.9
Channel x-section	Observations of river	7.9
Sediment load	Measuring turbidity and total solids	8.9
Channel alteration	Survey channel	7.9
Flooding frequency	Survey people	7.9
River habitat	Habitat assessment	7.12
Bank & riparian vegetation	Survey bank and riparian vegetation	7.7

Program Design

Introduction

The measurements and observations described in this manual taken together span the physical, biological, and chemical elements that define the ecological integrity of the river and its catchment. To the extent that any of these elements are degraded, they may affect the ability of aquatic life to live and reproduce, and may impair human usage of the river. Programs to assess rivers and their catchments are usually begun in response to a water quality or water quantity problem. Although effective programs follow similar paths in their development, each is unique and reflects the physical nature of the river, the surrounding land, and the people involved.

Rivers of the world and their catchments vary in many ways: size of catchment, hydrology of the river, surrounding land uses, human water usage, amount and type of riparian vegetation, and more. Readers of this manual, and participants in the GREEN network also vary in terms of resources available, degree of affiliation with established monitoring programs, the needs and concerns that drive monitoring efforts, understanding of the river and its catchment, and monitoring skill level.

The purpose of this chapter is:

➤ to describe a measurement and observation path that builds from an aesthetic (qualitative) understanding of the catchment and its river to more quantitative measures taken along a river reach; and,

➤ to describe the factors that shape the design of a water quality and catchment study and to offer guidance in building an appropriate design based upon these factors.

Assessment of a River and its Catchment

The path described here follows a general pattern beginning with the use of physical indicators (remote sensing and riparian surveys), leading to the use of biological indicators (benthic macroinvertebrate survey), and arriving at the use of physical-chemical and chemical indicators (dissolved oxygen tests, nitrates, and more—see Figure 5.1). Moving along this path, one develops a "big picture" view of water quality in its many dimensions, but also a deeper understanding of water quality at the local level. Those groups and individuals who are developing a river and catchment monitoring program can choose appropriate entry ways along this path depending upon the factors described later in this chapter.

Designing an Assessment Program

Designing an effective river and catchment assessment program requires understanding of two aspects: the physical nature of a river and its catchment; and the human dimension of the program participants. From this understanding, one can develop an assessment program that not only meets human needs, but that is also sensitive towards the physical character of the river.

Rivers around the world vary in size of catchment, hydrology of the river, surrounding land uses, human water usage, amount and type of riparian vegetation, and more. The design of effective river and catchment assessment programs requires attention to the physical nature of the river and its surroundings.

Many GREEN catchment and river assessment programs include three primary steps: preliminary assessment, assessment, and action-taking. Each step is built upon the next; the answers and questions generated by one step inform and shape the next. River and catchment assessments are based upon the following questions:

➤ Why assess water quality? What questions will we address?

➤ What do we want to assess? What indicators need to be measured or observed?

➤ How will we make those assessments? What approaches and methods should be used?

➤ Where do we want to make our measurements and observations?

➤ When do we want to make our measurements and observations?

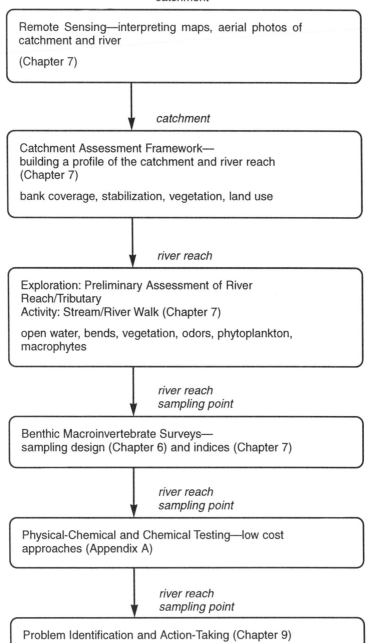

Figure 5.1. Assessment path.

The first two questions, "why" and "what," can be addressed through a preliminary assessment.

Preliminary Assessment

A preliminary assessment often includes looking at topographic maps or aerial photos to begin to identify potential sources or areas that contribute to water quality problems. Upriver or local land uses may provide a focus for the development of questions that need answers. "Is it safe for cattle to drink from this reach of the river?" "Is this village or community contributing raw sewage to the river and to the drinking water of downriver communities?" "Is erosion causing cloudiness or turbidity in the river?" "Why are fish or clams suddenly less plentiful in this river reach?"

Once potential assessment sites have been identified, a stream or river walk may be the next step. Activities in Chapter 7 describe many approaches to measuring and observing physical characteristics of bank and riparian vegetation, physical characteristics of the channel (discharge, slope, substrate), erosion, local land uses, odor and appearance of the water, and habitat assessment. From these activities, more specific questions will emerge. These questions will help determine which indicators to measure: phytoplankton, benthic macroinvertebrates, physical-chemical or chemical measures (see chart on next page).

For example, areas of eroded banks along the river may be causing higher turbidity downstream, which may lead to an assessment program that includes turbidity testing above and below this area. Benthic macroinvertebrate communities might be impacted if there is sedimentation downstream, so benthics become indicator organisms to measure (Chapters 6 and 7). What if a sulfur odor was noted on the river walk? This smell often accompanies areas of low dissolved oxygen, which may lead to an assessment program that includes dissolved oxygen and biochemical oxygen demand testing (Appendix A).

The table below links land uses with potential pollutants. Indicators are grouped as primary or secondary; a classification that helps to guide investigators to relevant measurements and observations. The initial listing of an indicator is tied to an activity from Chapters 7 and Appendix A (noted as "A" below).

Assessment

This step in the program is often a more organized and focused approach that builds from the questions addressed in the preliminary assessment and contributes to the "how", "where", and "when" questions above. The purpose of assessment is to determine, through measurements

Catchment Land Uses	Potential Pollutants	Primary Indicators	Secondary Indicators
Agricultural	erosion/sedimentation	turbidity (A) total solids (A)	benthics (5.1, 5.2, 5.3)
	pesticide runoff	benthics	
	nutrient runoff	nitrates (A) phosphates (A)	phytoplankton (4.1) macrophyte (4.2) dissolved oxygen (A) BOD (A)
	animal waste	fecal coliform (A)	nitrates phosphates BOD
	erosion/sedimentation	turbidity total solids	benthics
Residential	nutrient runoff	nitrates phosphates	phytoplankton macrophyte dissolved oxygen BOD
	human/pet waste	fecal coliform	nitrates phosphates BOD
	stormwater runoff	temperature (A) total solids turbidity	
Industrial	toxic discharges	benthics	
	thermal discharge	temperature	benthics phytoplankton dissolved oxygen
Mining	acid drainage	pH (A)	benthics
	heavy metals	benthics	
	erosion/sedimentation	turbidity total solids	benthics

and observations of indicators, whether there is a water quality problem, and the source of the problem. Given an understanding of the existence of the problem and its source, it is then possible to take effective action.

The assessment approaches chosen will be based upon the physical character of the river, the skill level of the group, others who might use the data or observations, and the resources available to the group. Most of these factors will be discussed under the "Human Dimensions of Assessment" later in the chapter. "Where" to make measurements and observations is based primarily upon the area defined in the preliminary assessment.

Figure 5.2. Students discussing the impact of water current on bank erosion.

"When" to make measurements and observations is influenced by the nature of the suspected water quality problem: it may be seasonal (e.g., during the rainy season); or it may be temporal, such as infrequent, but regular point source pollution.

For example, in much of the arid world, rivers are intermittent—full during infrequent heavy rains but dry much of the time. In such a river, benthic macroinvertebrate sampling would be ineffective because these organisms could not live in this habitat. Likewise, in very large rivers like the Amazon, Nile, Congo, Ganges or Yangtze, calculating discharge following the directions offered here (and with the resources available to most people) is unrealistic. In this case, tributaries probably account for much of the water quality anyway, and these smaller rivers may be easier to measure and observe. There may also be safety concerns from resident animals—like the crocodile along the Nile or in Australian rivers—which may influence how and at what times samples are taken.

Action-Taking

An assessment program often generates another set of questions that generate another round of measurements and observations that lead even closer to defining the source(s) of the problem. The questions addressed under the action-taking step include: What are the root causes

I Figure 5.3. Student reviewing her field notes.

of the problem (definition of the problem)? Who is affected by the problem? (people, animals). Who is responsible for this problem?

The transition from observations and measurements to interpretation of the data and definition of the problem is critical. Chapter 9 includes scenarios of students engaged in preliminary assessment, assessment activities, and action-taking.

Human Dimensions of Assessment

The following questions address the other important aspect, with the physical nature of the river, that helps determine the direction a water quality monitoring program will take: the human dimension.

Figure 5.4. Teachers learning how to monitor a river in a workshop.

> Who will use the surveys and water quality measurements? Do they have quality control needs that must be met?

> What are the resources available to the group?

> What is the general skill level in terms of taking measurements and interpreting observations?

> What is the level of experience with the river and its catchment?

The answers to these questions will help in planning a program that is at an appropriate scale and that seeks to answer the right questions. Consider these questions:

Who will use the surveys and water quality measurements? Do they have quality control needs that must be met?

Some river and catchment assessment programs may ask the kinds of questions that other government agencies and conservation groups are asking. In this case, perhaps it makes sense to find out what their water quality data needs are. Some data quality objectives may require equipment or techniques that are too expensive, or require too great a time commitment. Within the benthic macroinvertebrate surveys, phytoplankton surveys, types and amounts of macrophytes, catchment assessment surveys, and river and stream walks, there may be real opportunities for local citizens and programs to provide high quality data that the agencies do not have.

Formation of a steering committee: representatives from water agencies, nongovernmental organizations, schools, community leaders and planners.

Role: Program development, find needed resources, identify questions, and develop goals.

Goals of the program: to increase awareness about the river and catchment, to improve education through assessment activities and action-taking, and to improve the condition of the river.

Steering committee generates issues or questions that will influence the direction of assessment: most significant is untreated sewage entering the river. Where is the source of the problem? How large is the problem? What can this program do to solve this?

Physical character of the river and its catchment: preliminary assessment (using maps and aerial photos)

➤ maps indicate a large catchment, headwaters in mountains
➤ seasonal high flows; very muddy looking
➤ aerial photos show large agricultural areas, villages

Stream/River walk observations:

➤ odors, sulfur smell along river reach
➤ significant erosion along river bank—cattle?
➤ loss of natural riparian vegetation, much exotic "weeds"

Primary indicators: fecal coliform, turbidity

Assessment program (scope and nature of program based in part on resources and skill level)

Action-taking, see Chapter 9

Figure 5.5. Development of a Program.

What are the resources available to the group?

Resources, as used in this question, includes money to purchase equipment, support a coordinator, and provide for training. Resources could also include people in the community who can provide technical assistance, and institutions that will share equipment like nets, aerial photos, and testing kits. Some activities, such as the catchment assessment survey, stream/river walk, and even the benthic macroinvertebrate survey, can be performed with little or no resources.

What is the general skill level in terms of taking measurements and interpreting observations?

For some groups, skill level could be a limiting factor in the kinds of measurements or observations made. Generally, many of the measurements, observations, and activities described in Chapter 7 use only the three senses: touch, sight, and smell. Low-cost chemical monitoring (Chapter 8 and Appendix A) requires some skills in mixing chemicals, and performing chemical experiments.

What is the level of experience with the river and its catchment?

The level of experience one has with the river and surrounding lands clearly helps frame more specific questions that an assessment program could answer. Although one might be tempted to rush into fecal coliform monitoring to confirm a suspected point source of contamination, it is probably more effective in the long run for all groups—even those with high levels of experience with the river—to conduct a stream/river walk or catchment assessment survey. A walk or survey may uncover non-point sources of fecal coliform contamination to a point source, or may lead to other potential point sources and problems.

A Model Program

The model shown on the opposite page is developed from several successful GREEN river and catchment assessment programs that consider both an understanding of the physical nature of a river and its catchment, and the human dimensions of program participants.

An Ecosystem Approach

The physical, biological, and chemical character of the river and its catchment provide a comprehensive, ecoystem approach to understanding water quality. By taking the time to consider the questions offered in this chapter, investigators can design an assessment program that seeks answers to the right questions, and solutions to real problems.

CHAPTER

Benthic Macroinvertebrates

Introduction

Benthic macroinvertebrates are small bottom-dwelling organisms that can be seen with the unaided eye. Macroinvertebrates form an integral biological component of river ecosystems and often reflect the qualitative character of flowing waters. Their importance to humans is both direct and indirect. Many types of biting flies such as mosquitoes, black flies and horse flies that can pass diseases on to humans and their livestock have life stages that occur in flowing waters. Some larger macroinvertebrates, such as mussels and crayfish, are often used either as food or as bait to obtain fish. Indirectly, many forms of macroinvertebrates play an integral part in river ecosystem health. They are often the most important organisms linking primary producers like algae and aquatic macrophytes and other organisms which depend on macroinvertebrates for food, particularly fish.

Rivers contain a large diversity of macroinvertebrate organisms; only the groups which will be considered for water quality monitoring will be described. Other forms are ecologically significant, but may be too small or are otherwise difficult to effectively sample.

Important groups include flat worms (Turbellaria), tubificid worms or oligochaetes (Oligochaeta), leeches (Hirundinea), sowbugs (Isopoda), amphipods (Amphipoda), crayfish (Decapoda), snails (Gastropoda), clams and mussels (Pelecypoda), and insects (Hexapoda = Insecta). Although insects are primarily terrestrial, some orders, or groups within certain orders, have become adapted to aquatic life during some part of their life history. Several insects groups are frequently used as bioindicators of water quality. Insect orders which we shall examine include the mayflies (Ephemeroptera), dragon- and damselflies (Odonata), stoneflies (Plecoptera), dobson- and alderflies (Megaloptera), spongillaflies (Neuroptera), caddisflies (Trichoptera), beetles (Coleoptera), true bugs (Hemiptera), and true flies (Diptera).

Macroinvertebrates are widely used as indicators of water quality, and many indices have been developed to evaluate water quality based on the

Figure 6.1. Stonefly nymphs are indicators of good water quality because they cannot tolerate low oxygen, siltation, warm water, or organic enrichment.

presence or absence of certain taxa. These indices assume that polluted sites or systems generally contain fewer taxa than unimpacted ones and that relative absence or presence of certain species is a direct result of water quality conditions.

Macroinvertebrates and Other Bioindicators of Water Quality

Physical and chemical tests (Chapter 8) yield data on present water quality. However, such data may not show the effects of past conditions, and may not predict how biological organisms will respond to conditions over time (Plafkin et al. 1989).

Although many organisms can be used to monitor water quality, the "ideal" characteristics that bioindicators should possess are: taxonomic soundness and easy recognition; broad distribution to facilitate application to other regions; abundance to permit easy and repeatable sampling; large body size to facilitate sampling and sorting; limited mobility and relatively long life history; and available data on organism ecology (Johnson et al. 1993).

Fish are frequently used to monitor water quality because their biology and food and habitat requirements are often well known, and because water quality standards, legislative mandates, and public opinion are often directly related to the value of a waterbody as a fishery resource (Plafkin et al. 1989). Fish are good indicators of long-term effects and broad habitat conditions, and, because they are often directly consumed by humans, their health may affect our own. However, thorough sampling is difficult and low population densities make reliable interpretation of sample data difficult. Also, fish may be absent from certain habitats, such as very small streams. Because fish are comparatively long-lived, it may take some time for changing water quality conditions to have noticeable effects. Finally, proper sampling techniques, such as electro-shockers or seines, are expensive and dangerous for inexperienced users, and often result in widespread instream habitat disturbance. This manual will not discuss sampling fish to monitor water quality.

Rapid reproduction and short life-cycle make algae good short-term indicators of water quality. As primary producers, algae are most directly affected by physical and chemical disturbances. Also, algae are easily sampled with minimal impact on stream habitats. Relatively standard methods exist for detecting effects of pollution such as toxic substances by measuring biomass and chlorophylla, or by identifying pollution tolerant algae (Weitzel 1979; Rodgers et al. 1979). However, algae undergo dynamic and short-term natural cycles of change in population (Hoagland et al. 1982). Also, taxonomic keys may not be readily available, and high-powered (400x) equipment is often required to make identifications. Inexperienced investigators may find some algal groups, such as diatoms, difficult to identify. Finally, several kinds of blue-green algae are highly toxic, and therefore only those experienced should make such an analysis. An activity for sampling algae is included in Chapter 7.

Macroinvertebrates provide an excellent compromise between the two other main groups of organisms. They are good indicators of local conditions and site-specific impacts. They quickly integrate the effects of short-term environmental variations. Because they are not very mobile, they are easily sampled with inexpensive equipment with minimal disturbance to stream habitats. Also macro-invertebrates are abundant in most streams, and many smaller streams may naturally support a diverse population of macroinvertebrates but only a few or no fish. Macroinvertebrates serve as a primary food source for many fish, making them an important fisheries resource. Finally, although perhaps not as easy to identify as fish, with some training, macroinvertebrates are not difficult to identify, especially with easily obtainable magnifying glasses or small microscopes.

Limitations

There are some drawbacks in using macroinvertebrates to assess water quality. Large numbers of organisms must be sampled, and the absence of any major group of organism may not reflect water quality conditions. Most available indices largely apply to temperate zones of North America, Europe, and Australia. Adjustments must be made to account for differences in ecosystems and biogeographical distributions. For example, stoneflies will most likely not be found in warm, slow-moving tropical streams simply because they are not found in that habitat. Certain taxa naturally do not occur in downstream areas where silt and clays are deposited. Our taxonomic and biological knowledge is still very incomplete in many regions of the world, particularly South America, Africa and Asia. There also may be a large degree of variability within a group of organisms with respect to tolerance to pollution or degraded water conditions. For example, fly larvae (Diptera) include very pollution-tolerant groups (moth flies-Psychodidae, blood-red midges Chironomidae) as well as very intolerant families (net-winged midges—Blephariceridae; snipe flies—Anthericidae). Indices also do not incorporate naturally-occurring ecological pressures on macroinvertebrates, such as floods and fires, or predation from fish. Indices are estimates based on sampling protocols, and, whenever possible, should be used in conjunction with other tests (e.g., dissolved oxygen, turbidity, fecal coliform, etc.) to form conclusions about water quality.

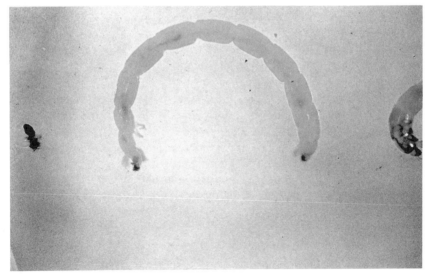

Figure 6.2. Midges may be abundant in polluted streams. Some midge species possess hemoglobin, allowing them to live with low levels of dissolved oxygen.

While acknowledging these limitations, macroinvertebrate sampling yields valuable data on water quality conditions and helps pinpoint sources of degradation, as well as providing a fascinating look at the small organisms that share the Earth's waters.

Macroinvertebrate Sampling

Achieving a representative benthic sample in order to assess water quality can be a difficult task. One must consider the complexity and variety of habitats found in streams. Each habitat can support large numbers of different types of organisms, which change in form and size. Even a complete sample of a limited river reach may not represent the overall water quality of the river. Consider a small reach of stream substrate with patches of algae amid cobble and silt, near a grassy bank with some woody debris. The larvae of stoneflies, mayflies, net-spinning and predacious caddisflies, blackflies and some midges as well as snails, attached clams, and leeches may co-inhabitant the same rock. In the silt or gravel in which the rock is partially embedded, one may find dragonfly or damselfly larvae together with water pennies, riffle beetles, midges, worms, and amphipods. The substrate may contain water bears, burrowing mayflies, sowbugs, and other organisms using the small spaces for refuge, food, and oxygen. The grassy banks shelter water scorpions, and on the water's edge one finds back swimmers, giant water bugs, water treaders and striders, and other species. Organisms may be found on or even in wood debris, and completely different animals are found in areas of leafy accumulation.

Another reach of stream may contain only a few or perhaps no readily identifiable benthic organisms. There are many reasons for this. Water quality may be poor, with degraded habitat quality or quantity. The scarcity of organisms may be temporary, such as when insects are in their adult and largely terrestrial stage, or in an egg stage when their presence is not readily noticed to the unaided eye, or following a severe flood. Animal scarcity may be the natural state, such as in silt and sand bars of very large rivers, which usually support only specialized mayflies and dragonflies, along with oligochaete worms and clams (Williams and Feltmate 1992). Human error or disturbance during sampling may cause an apparent scarcity. For example, improper use or using the wrong sampling device may cause investigators to miss organisms. Excessive disturbance of stream substrates by over-eager collectors may cause many organisms to drift away. It is always a good rule to limit one's activity in the stream, disturbing this ecosystem as little as possible. Finally, animal scarcity may indicate a more serious condition, caused by pollution or other human impact within the catchment.

Figure 6.3. Students using home-made nets to sample their monitoring site for benthic macroin-vertebrates.

It is important to have reference data or "benchmarks" on the types of macroinvertebrates which naturally occur in the stream of interest to compare to samples from sites which are potentially affected by pollution or habitat degradation. This is necessary in order to calibrate your macroinvertebrate index. Unfortunately, many streams have not been sampled in the past. However, there are several ways to acquire this information depending upon your sampling protocol.

Figure 6.4. A Surber sampler is used to collect quantitative data in order to compare a control site with their school's monitoring site.

Sampling Protocol

Point Source Pollution

Sampling macroinvertebrates to determine the effects of point-source pollution, usually requires no more than sampling an upstream control location and comparing that sample from downstream treatment locations. Samples from the control site should be far enough upstream from the point-source location that no bias is introduced. A possible sampling strategy would be to sample a distance of 100-300 m upstream for the control site and a safe distance downstream from the point-source pollution site. More than one treatment site might be sampled in order to determine the magnitude of the point source disturbance upon the benthic environment downstream. One can also pinpoint when pollution effects the stream by sampling both before and after the occurrence of point-source pollution and identifying differences between the two times frames.

Nonpoint Source Pollution

Sampling nonpoint-source pollution or other types of disturbances that effect large river reaches, such as sedimentation from farmland erosion, widespread logging and removal of riparian vegetation, and application of herbicides or pesticides over fields and trees, is more difficult because there is no clearly defined reference point. A "paired" sampling approach can be used by which an undisturbed (control) stream is compared with the affected (treatment) stream. Similar-sized streams in the same area or ecosystem are likely to share similar fauna, so that observed differences in fauna and flora between the two rivers are due largely to differing water quality or physical conditions.

To monitor macroinvertebrates using the paired sampling approach, first locate your stream on a catchment map and determine stream order and measure discharge on-site. Now, from your map and field observations, choose a stream that shows considerably less evidence of stream degradation but shares similar physical characteristics (discharge and drainage area, width, depth, etc.). Aquatic macroinvertebrate samples should be taken at three locations approximately the same distance apart on each stream, following the habitat-sampling method described above. Compare taxa from your control stream with that of your treatment system. Remember to first ascertain hazards involved in sampling treatment sites - use only the safest sampling methods possible.

Benthic Sampling and Devices

Thorough qualitative and semi-quantitative sampling may require identifying and sampling distinct river habitats, including riffles, pools, stream banks, plants, rocks and boulders, and the substrate. Generally, qualitative sampling merely attempts to identify as many different types (taxa) of organisms as possible that can be found in the stream. Water quality is indicated by comparing identified animals with a list of those expected in those habitats. Semi-quantitative methods obtain information about diversity and relative abundance of various organisms, which can also yield useful information regarding water quality. These latter methods, however, require more rigorous methods, such as a prescribed number of samples, randomly selected stream areas, and a fixed duration in sampling.

Kick Seine

Perhaps the easiest benthic sampling device to construct is a hand screen, which can be used to collect organisms dislodged from stream substrates through kicking or "kick seining" (Figure 6.5). Kick seining is one of several standard methods for qualitative and semi-quantitative sampling, particularly of riffles (e.g. Plafkin et al. 1989). Kick seines can be elaborate devices with precise 250 um mesh screen with weighted bottom edges to prevent dislodging by current or bedload, or relatively simple devices made

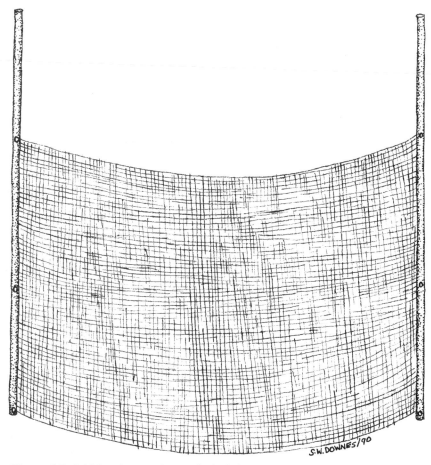

Figure 6.5. A kick seine can be easily built by a class and used to sample benthic macroinvertebrates. (Drawing by S.W. Downes.)

with window screening attached to two pieces of wood (see Appendix A). All that is required are screen holes small enough to collect small organisms but large enough to permit water to pass through without resistance (<1 mm is recommended). Kicking the substrate on the current (upstream) side of the screen for about 30 seconds is usually sufficient to dislodge organisms attached to stones or buried in the substrate. Semi-quantitative procedures using kick seining often require a fixed area—(usually 1 m^2) be sampled for a timed period—usually one minute. In this way, comparisons between samples can be made without introducing bias. Kick seining works best in areas with adequate flow whereby dislodged organisms will be washed onto the screen and retained when the seine is removed from the sampling position. This method may seem simple, although certain techniques are required when working in fast-flowing waters or when heavy debris (wet leaves, stones) have collected on the screen.

D-frame Nets

D-frame and other aquatic nets (Figure 6.6) are relatively easy and inexpensive to construct. They are very useful for sampling riffles and runs, pools, drifting and surface organisms, aquatic vegetation, and overhanging grasses and bushes. The long handle makes the D-frame easy and safe to use, and allows collecting in dangerous or difficult-to-reach places such as deep pools, log jams, and under thick overhanging brush. In addition, aquatic nets are useful for sampling slow-moving river bottoms or lakes, where there is often little current to wash organisms onto screens. Semi-quantitative methods often prescribe a fixed number of net sweeps through vegetation, along the substrate, and within the water column. One can also combine kick seining techniques with nets in order to avoid the tendency of screens to collect heavy stones and awkward woody material when dragging along the substrate or sampling near the shore. The D-frame net is perhaps the most widely used qualitative sampling device for both lotic and benthic systems.

Artificial Substrates

Good results can be achieved with various types of artificial substrate samplers, such as basket samplers or trays filled with stones, gravel, or leaves, multiple-plate (e.g. Hester-Dendy, Figure 6.7), leaf packs (usually collected riparian leaves tied together), or simple surface devices such as bricks or shingles. Substrate samplers are very useful in streams where suit-

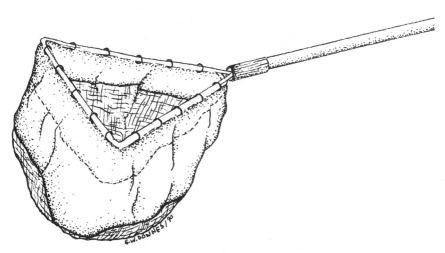

Figure 6.6. D-frame net. (Drawing by S.W. Downes.)

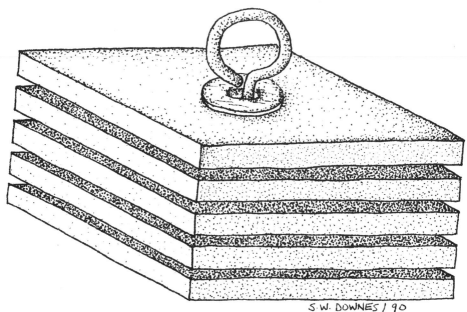

S.W. DOWNES / 90

Figure 6.7. Hester-Dendy artificial substrate sampler (homemade). (Drawing by S.W. Downes.)

able sampling habitat is difficult to reach or find, or when one wishes to minimize physical contact with water that may be unhealthy. Drifting organisms will colonize these substrates over a period of time and allow the investigator to collect organisms that may not normally be collected by kick seining or net devices. Artificial substrates have a distinct advantage in semi-quantitative sampling, because the sampling location can be precisely controlled with regard to area, number of plates, placement, etc. However, there are some disadvantages to "passive" collecting. A lengthy time (up to eight weeks) may be required before sufficient quantities of representative organisms colonize the substrate. Adequate space must be left between stones or plates to allow larger species to colonize. Successful colonization will only occur on substrates that accumulate enough food, such as algae, or that are positioned in a current that brings food to sessile organisms. Leaf packs must be broken down, or "preconditioned" by microorganisms (bacteria and fungi) so that organisms can feed on this material (Williams and Feltmate 1992). Finally, substrate samplers can be dislodged by flooding, covered by silt, or damaged or destroyed by people. To prevent the latter, substrates must be carefully hidden, and the investigator should make either discrete marks nearby to indicate their location, or carefully note their position relative to a fixed point on the bank.

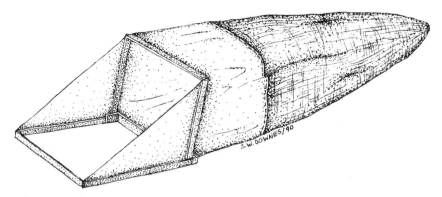

Figure 6.8. Surber sampler. (Drawing by S.W. Downes.)

Surber Samplers

Another quantitative sampling method uses Surber samplers (Figure 6.8). These devices are fixed-area samplers placed upon the stream substrate. The researchers disrupt the substrate within the sample area with their hands, rubbing attached organisms off larger rocks, and digging up the stream bed. Organisms are carried by the current into net attached to one end of the sampler. Care must be taken not to get too much debris into the collection net. These devices make capturing macroinvertebrates in shallow streams quite easy. However, they are difficult to construct and often expensive to purchase. Also, they are difficult to use in deeper river reaches, and may not be appropriate for young collectors.

Hand Collection

Hand collection involves collecting rocks, wood, and other large objects, and simply picking or rubbing organisms into a tray, net or other collection container. This method is often used in conjunction with other sampling procedures as a way to collect organisms not easily captured with nets, screens, or other sampling methods. Certain organisms, particularly those found in specific habitats such as moss and wood, or attached to large rocks or boulders, can only be hand-picked. However, this method is very time consuming. Take care when sampling to avoid sharp or pointed objects, such as wood pieces with splintered edges concealed by algae or silt, or rocks with jagged edges. A plastic table knife or other object can be used to gently scrape surfaces.

Many of the above methods can be used to sample the margins of larger rivers. However, depth and current makes mid-sections of these rivers difficult and often dangerous to sample with these methods. In large rivers with riffles, the benthos may be safely sampled by using basket or artificial substrate samplers.

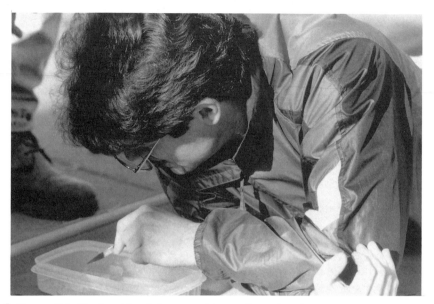

Figure 6.9. Student identifying organisms collected and placed in a container for identification.

Sampling Adult Insects

The larvae of insects often receive the most attention by collectors. However, it is also very easy to collect and identify many of the adult forms of insects with aquatic stages. Many flying insects are positively phototrophic, that is, attracted to light. The Mundie pyramid trap is frequently used to collect adults as they emerge from streams. Other light traps can be constructed and placed next to the stream to collect various adult species. Light traps are useful during seasons when many insect species are in the adult stage, and cannot be sampled in the stream. Also, additional aquatic insects may appear that were not observed during aquatic sampling. Light traps are easy to construct but require a bright light source and a power source, such as a battery or generator. Also, the investigator must identify and eliminate the terrestrial species which are not of interest.

Macroinvertebrate Indices

The health of the river ecosystem can be assessed using numerous biotic indices which utilize various levels of taxonomy. More precise indices make distinctions at the level of genera or species, but, because these are often time consuming and require high powered microscopes for identification, this manual will rely on indices that use more general levels of taxonomic distinction. Species- and generic-level distinctions have been shown to be more predictive of stream community (Hilsenhoff 1988) but family-level indices are also comparable (Lenat 1988; Resh and Jackson 1993). Depending on the taxonomic expertise, resources (time and equipment), and taxonomic keys available, investigators may choose a group index (i.e., based on order of easily identifiable organisms), or family/genus/species-based protocol.

Limitations of Indices

As described earlier in this chapter, some macroinvertebrate groups display predictable pollution tolerance. Stoneflies and mayflies require high water quality. Mussels, crayfish, caddisflies and dobsonflies, water mites, and dragonflies require good water quality, whereas damselflies, beetles and snails are often found in waters of fair quality. Blood midges and aquatic worms often predominate in waters of very poor quality. However, because many taxa vary considerably in their tolerance to various types of pollution (organic enrichment, sedimentation, etc.), it is difficult to devise a water quality index that can be consistently reliable in all part of the world.

There are several reasons for this limitation. First, many groups of macroinvertebrates display large ranges in tolerance to pollution. Stream reaches directly affected by industrial pollution are usually characterized by high densities of midges and oligochaetes and an absence of mayflies and stoneflies (Williams and Feltmate 1992). However, some mayfly species can tolerate moderate levels of organic pollution (see Hilsenhoff 1988). Most indices that require identification to family, genus, and species take into

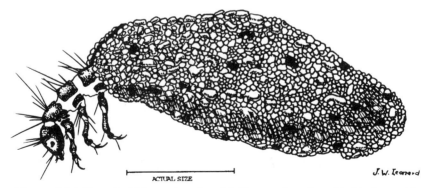

ACTUAL SIZE

J. W. Leonard

Figure 6.10. Over 10,000 species of caddisflies are found worldwide (except in Antarctica). Many kinds of fish feed on caddisfly larva, pupa, and adults. (Drawing by J.W. Leonard.)

account the variation that exists among macroinvertebrate orders. Also, any particular group of organisms may vary considerably by region in their tolerance to pollution. For example, Australian indices rate isopods (Asellus) as indicators of good water quality, whereas European and North American indices rate them tolerant of considerably lower water quality.

Second, many areas of the world do not have their macroinvertebrate fauna identified and documented. This situation is now improving with more international communication and access to informational databases. Third, identification and known distribution of various types of taxa, especially at the level of genus and species, may be poor, due to the fact that many streams and river reaches have yet to be sampled. This too is improving with increasing interest and research in lesser known river systems of the world. Perhaps water quality monitoring programs initiated by this manual will contribute to our knowledge of benthic macroinvertebrates.

Limitations in the tools and methods used to evaluate macroinvertebrates should be recognized when collecting samples. For example, timing of sampling must take into account macroinvertebrate life history and physical river conditions. Macroinvertebrates that normally occur in a river will not be observed when they are in the egg or adult life-stage, if sampling protocols only analyze larval stages. Summer sampling may yield different results than sampling during fall or early spring due to life histories and flow conditions affecting macroinvertebrate distributions. Proper sampling techniques will help limit the bias that may occur during collection. Finally, recognize the limitations involved in the absolute accuracy of the biotic indices used to make inferences from data.

Scores and Analysis

Developing one particular scoring system for many regions of the world is a challenge. However, comparison of indices from various parts of the world do reveal some generalizations from which to derive water quality scores. Most indices assign tolerance ratings for each organism to water quality. For these activities, we will assume that the level of taxonomic experience is minimal, and the ability of recognizing insects is limited to order and perhaps some families. However, if you have access to an index specifically developed for your area or region, or you have greater taxonomic experience, your results may be more accurate and reliable.

Streamwatch and Waterwatch Programs (Australia)

These programs evaluate water quality based on a composite score of different types of macroinvertebrates in a sample. The indices are very easy to understand and difficult math is not required. Comparisons can be made

Figure 6.11. Students sample their monitoring site at the same time each year.

among sites within a stream, or between different streams. The types of organisms used in the index are relatively common and easy to identify. This has the ecological advantage that identifications can often be made rapidly, so the organisms can be returned to the stream unharmed. Both systems rank groups of organisms into tolerance classes, rating organisms in terms of their tolerance to pollution. Although Streamwatch uses a scale of 1 to 8, and Waterwatch a scale of 1 to 4, both indices agree on the relative sensitivity of particular organisms to pollution, with 1 indicating the most pollution-tolerant organisms.

To follow the Streamwatch method, identify each different type of organism from a fairly representative benthic sample of all important river habitats, and the pollution-tolerance value for each type, and then add the values for a stream pollution index. The cumulative score yields the water quality rating: 20 or less = poor, 21-35 = fair, 36-50 = good; 51 or more = excellent. For example, a benthic sample yields the following organisms (pollution tolerance rating): mayflies (7), crayfish (7), dragonflies (6), mussels (6), shrimps (6), beetles (5), leeches (3), snails (3), mosquitoes (1), and blood worms (1). Values in the parentheses sum to 45, which yields a "good" stream quality rating.

Rather than using a cumulative index, the Waterwatch program uses a sequential comparison index (SCI), which assumes that a healthy and diverse aquatic community should yield a high habitat-qualitative score (see also Cairns and Dickson, 1971). This formula yields a 0.0 to 1.0 scale of water quality where a value of 0.0-0.3 is rated as poor water quality, 0.3-0.6 is fair, 0.6-0.8 is good, and 0.8-1.0 is excellent.

Description of Taxonomic Groups

The following is a brief description of the major taxonomic groups of benthic macroinvertebrates that may be found in a stream or river. (See Chapter 7 for a listing of distinguishing characteristics of common taxa.)

STREAMWATCH			WATERWATCH		
Organism	Rating	Ranking	Organism	Rating	Ranking
Stonefly	8	Very sensitive	Stonefly	4	Excellent
Mayfly	7	Very sensitive	Mayfly	4	Excellent
Mussel	6	Sensitive	Mussel	3	Good
Shrimp, Fr. W	6	Sensitive	Shrimp, Fr.W	4	Excellent
Dragonfly	6	Sensitive	Dragonfly	3	Good
Caddisfly	6	Sensitive	Caddisfly	3	Good
Amphipod	5	Sensitive	Amphipod	3	Good
Water mite	5	Sensitive	Water mite	3	Good
Beetle larvae	4	Tolerant	Beetle larvae	2	Fair
True bugs	4	Tolerant	True bugs	2	Fair
Snails	3	Tolerant	Snails	2	Fair
Leech	3	Tolerant	Leech	2	Fair
Flatworm	3	Tolerant	Flatworm	2	Fair
Fly larvae	2	Very Tolerant	Fly larvae	1	Poor
Midge larvae	2	Very Tolerant	Midge larvae	1	Poor
Mosquito larvae	1	Very Tolerant	Mosquito larvae	1	Poor

Figure 6.12. Relative pollution sensitivity of particular benthic organisms using two different scales.

Flatworms

Only a few species of flatworms (Phylum Platyhelminthes: Class Turbellaria) inhabit freshwater habitats. They are elongated organisms that have world-wide distribution. Most species are very small (<1 mm), some are 5 to 30 mm long. Flatworms are found in most types of aquatic environments with sufficient food supply, typically under objects or in debris during the daytime (Pennak 1989). Although this group of organisms shows large variations in habitat tolerances, some species in the genus Phaenocora produce hemoglobin that enables survival in low oxygen sediments (Kolasa 1991).

Leeches

Leeches (Phylum Annelida: Class Hirudinea or Superfamily Hirudinoidea) are segmented, flattened animals characterized by a mouth which contains an oral sucker. Most species are scavengers or predators, but some species are parasites on warm-blooded animals. Although bloodsucking leeches attach to human skin, others feed on midges, worms, and other invertebrates, and are themselves prey for many terrestrial and aquatic organisms. Some are raised for fish bait. Most common in warm, still, protected shallow water, leeches avoid light and generally hide and attach themselves under concealing objects such as stones, plants, and wood debris (Pennak 1989). Most species inhabit freshwater habitats, though some are marine and terrestrial. Leeches are found in most geographical areas of the world.

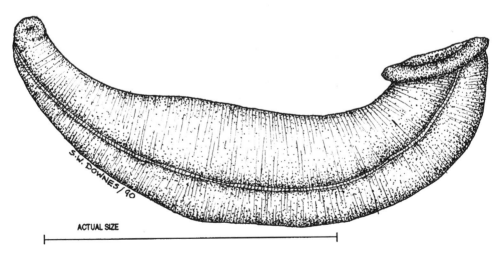

ACTUAL SIZE

Figure 6.13. Leech. (Drawing by S.W. Downes.)

Oligochaetes (tubificids)

Oligochaetes or Tubificids (Phylum Annelida: Class Oligochaeta) are found in every kind of aquatic habitat around the world, and display great taxonomic and ecological diversity. Oligochaete worms are bilaterally symmetrical and segmented (aquatic earthworms), and feed on substrate particulate organic matter, algae, and bacteria. Oligochaetes are adapted to live in a variety of substrate habitats, from sand to mud. The family Tubificidae, particularly the widespread species *Tubifex tubifex*, is known for its ability to develop dense colonies in organically polluted waters (Pennak 1978; Brinkhurst and Gelder 1991).

Aquatic Sowbug

Sowbugs (Phylum Anthropoda, Class Crustacea, Order Isopoda) are largely marine and terrestrial, but some species are found in smaller streams and lakes, where they scavenge dead animals and plant material (detritus). Sowbugs are identified by a dorsoventrically (top to bottom) flattened body, with pleopods (abdominal appendages) forming broad plates, and 6 or 7 pairs of legs. They are found under stones or in detritus. Some species are indicators of organic pollution, and can be abundant in areas downstream from streams polluted by domestic sewage (Pennak 1989).

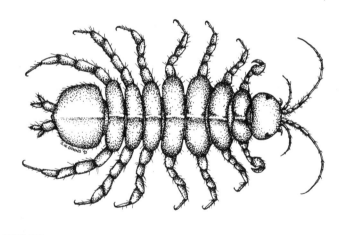

ACTUAL SIZE

Figure 6.14. Aquatic Sowbug. (Drawing by S.W. Downes.)

Amphipods

Amphipods or scuds (Phylum Anthropoda, Class Crustacea, Order Amphipoda) are a largely marine group with freshwater representatives that occur in a wide variety of unpolluted freshwater habitats. Most are 5–20 mm long, with a laterally (side to side) compressed segmented body. They are found in the substrate, where they feed on organic matter, algae, and freshly dead animals. Scuds are widespread on all continents where waters tend to be cold (except Antarctica). A large number of species have adapted to underground stream and cave environments.

Crayfish

Although the vast majority of Decapods are marine, crayfish (Phylum Anthropoda, Class Crustacea, Order Decapoda) are often an important member of freshwater river communities. Crayfish are characterized by a cylindrical body, stalked and movable eyes, and long antennae. The first five sets of appendages are often modified for handling and mincing food, sometimes with large chelae or claws. The hind five appendages (pereipods) are used for movement. Because many species are large and long-lived, they process large amounts of organic matter and thus play an important role in stream ecosystems. Found in many parts of the world, crayfish are particularly abundant in North America, but absent from Africa.

Mollusks (snails, clams, and mussels)

Mollusks are found in freshwater environments around the world. Most freshwater environments support snails (Class Bivalva, Order Gastropoda), which are characterized by their spiral or coiled shells. They feed on detritus and algal growths on substrates (thus are found more frequently in shallow environments), and can be important prey for fish. Most snails are limited to waters that have sufficient mineral content for shell formation, but a few species are found in low-carbonate waters. Snails are generally not found in waters with a pH of 6.2 or lower, because of their shell mineralization requirements (Pennak 1989). Two main groups in freshwater are the air-breathers (Subclass Pulmonata) and the water-breathers (Subclass Streptoneura). Severe anoxic conditions also are limiting, but certain species can survive particularly low DO levels and thus are found in areas with some organic pollution. Pulmonate snails, which breathe air, are quick to take advantage of organic enrichment. As pollution eliminates predators, pulmonate snails (e.g., Physa, Lymnea) thrive. Some snails are also the intermediate host for fluke parasites, and thus of human health importance.

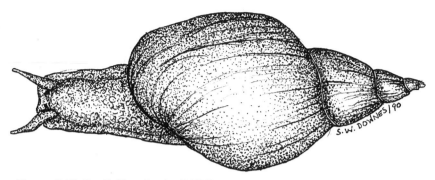

Figure 6.15. Snail. (Drawing by S.W. Downes.)

Clams and mussels (Class Bivalva, Order Pelecypoda) are protected by two symmetrical, opposing valves or shells, and feed by straining particles through two sets of lace-like gills. Clams and mussels need to cling to relatively clean substrates, and are intolerant of polluted water, excessive turbidity, and sedimentation. Many species are disappearing in many areas of North America, where the freshwater clams reach their highest biodiversity. Other species, such as Sphaerium, can tolerate considerable organic pollution. Clams are most abundant and diverse in large, shallow rivers.

Insects

Insects (Phylum Anthropoda, Class Hexapoda) are found in almost every freshwater aquatic environment on Earth, and have 6 legs on the thorax (as adults); a three-part body including the head, thorax, and abdomen; and a chitinous exoskeleton (see Figure 6.16). Insects undergo metamorphosis, a complex life cycle with several stages. There is, however, much life cycle variation within the insect group. Complete metamorphosis, is characterized by egg, larva, pupa, and adult. In this type of metamorphosis there are significant physical differences between adults and juveniles. Dobsonflies and alderflies, caddisflies, beetles, and true flies have this type of life cycle. Some undergo incomplete or simple metamorphosis in which the adult and juvenile are physically similar, and the second stage is called a nymph. Mayflies, stoneflies, and dragonflies are insects with this type of life history. Other groups, such as the true bugs, have winged adults and non-winged juveniles, but otherwise undergo a gradual transition.

There are 32 orders of insects, of which 13 live in the water during one or more life stages. Another 3 orders are associated with water habitats (semi-aquatic). Adults of all semi-aquatic and aquatic insects breathe atmospheric oxygen, although the methods of obtaining oxygen often do not require leaving the water.

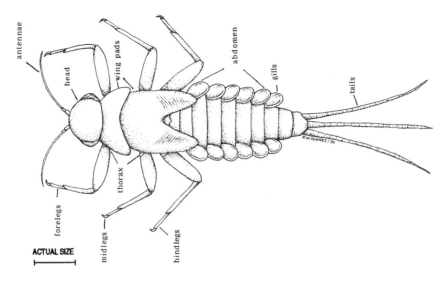

Figure 6.16. Mayfly, illustrating major body parts. (Drawing by S.W. Downes.)

Mayflies

Mayflies (Order Ephemeroptera) generally live in unpolluted water conditions, but some (Baetis, *Hexagenia bilineata*) are tolerant of some pollution and siltation. Nymphs are usually scrapers (algae) and collectors (fine detritus), though some species are specialized filter feeders or carnivores. Adults do not feed, and are short-lived (1-3 days). Mayflies, found in Permian and Upper Carboniferous sediments, are the oldest and most primitive winged insect with fossil species. This group is represented by 2100+ species in about 310 genera and 22 families. Mayflies can be found around the world, except for the high Arctic and Antarctic.

Dragonflies and Damselflies

Although the most diverse fauna of dragonflies and damselflies (Order Odonata) inhabit still or slow-moving waters, nymphs do occupy an array of running and standing inland waters. The three suborders, Zygoptera (damselflies), Anisoptera (dragonflies), and Anisozygoptera (2 species in Japan and the Himalayas) contain about 5500 species. Biodiversity increases strongly toward the tropics. They are strong fliers, enabling them to disperse widely. Both nymphs and adults are carnivorous. Aquatic nymphs feed primarily on small benthic animals, but include larger invertebrates and immature vertebrates (fish and amphibians) as they grow. They may have aquatic life stages lasting one year in warmer climates and up to five years in higher latitudes. Terrestrial adults live about 3 weeks and feed on insects.

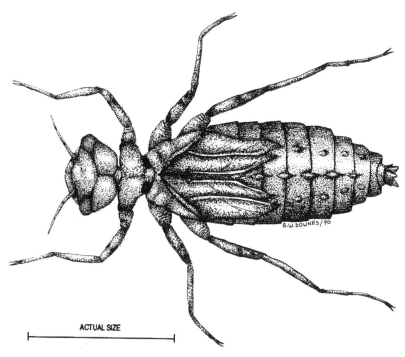

ACTUAL SIZE

Figure 6.17. Dragonfly nymph. (Drawing by S.W. Downes.)

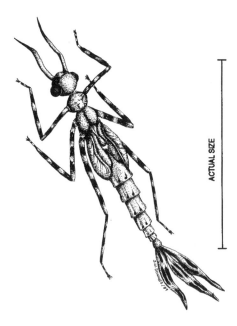

ACTUAL SIZE

Figure 6.18. Damselfly nymph. (Drawing by S.w. Downes.)

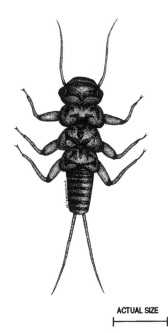

Figure 6.19. Stonefly nymph. (Drawing by S.W. Downes.)

ACTUAL SIZE
├────────┤

Stoneflies

Stoneflies (Order Plecoptera) are usually found in cool, clear streams, and are restricted to higher altitudes, circumpolar waterbodies, or spring-fed lowland streams. Stoneflies are usually considered indicators of good water quality because most cannot tolerate low oxygen levels, siltation, high temperatures, or organic enrichment. Nymphs of many species are associated with particular reaches of a stream, depending on the micro-habitat substrate, current regime, presence of other organisms, and local variations in water chemistry and temperature. About 2000 species are grouped into 2 suborders and 15 families. Nymphs of most families are primarily detritivores, though later nymph stages and larger species of some families are largely carnivorous (e.g. Perlodidae, Perlidae, Chloroperlidae, Eustheniidae).

True Bugs

True bugs (Order Hemiptera) are distinguished by nymphs that are very similar in appearance to adults, with similar habitat preference and behaviors, but are smaller (Williams and Feltmate 1992). There are about 3300 aquatic or semi-aquatic species grouped in 69 genera and 16 families.

Although a physically and ecologically diverse group, Hemipterans are characterized by piercing-sucking mouth parts, with a segmented rostrum or beak, and a forewing consisting of a membranous tip. Most are predacious, although one group—Corixidae—have many species that feed on detritus. They occupy many aquatic habitats, with most species found on water surfaces or margins of ponds, slow streams, or littoral zones of lakes (Ward 1992). Rapid streams have few species; normally on the water surface along the edges.

Dobsonflies (Alderflies)

Dobsonflies and Alderflies (Order Megaloptera) have just two families, Corydalidae (Fish- and Dobsonflies) and Sialidae (Alderflies). Only the larvae are aquatic; these are predaceous, and are common in clear river and lake habitats. Corydalidae are widely distributed throughout temperate regions (absent from Europe), with a few species found in the tropics. Larvae of the Sialidae are widely distributed but restricted to northern and southern temperate zones. The larvae tend to live in more turbid waters, or at least those with silty or muddy substrates, and prey upon smaller insects. Five to ten percent of the world megalopteran fauna occur in Australia.

I **Figure 6.20.** Dobsonfly.

Caddisflies

Caddisfly (Order Trichoptera) larva and pupa live in most types of waterbodies, both warm (up to 34°C) and cold, in springs, and temporary ponds. Tolerance to pollution varies widely, with some species being quite tolerant. Their abundance in most streams makes them a valuable fisheries resource. Larvae of many species build cases from silk. Adults live a few days to several weeks, and many are strong fliers that can disperse widely. Over 10,000 species are distributed worldwide (except Antarctica); some are restricted to either the northern or southern hemisphere. Although diverse in tropical regions, the greatest diversity is in cool running waters. Caddisflies feed on a wide range of foods. For example, many limnephilids consume detritus, diatoms, and other algae scraped or collected from substrate surfaces, whereas others are more selective, shredding riparian leaves. Many fish feed on caddisfly larva, pupa, and adults.

Beetles

Beetles (Order Coleoptera) are the largest known order of insects, and are mostly terrestrial. However, there are about 5,000 semiaquatic and aquatic species separated into 15 diverse families, of which 10 have adult and/or larval stages in aquatic environments. Adult beetles are characterized by heavy exoskeletons. Larvae are physically variable, but generally

I Figure 6.21. Caddisfly.

have hardened head capsules. Because adults disperse widely, beetles have worldwide distribution. Unlike other aquatic insects, some species of beetles are fairly long-lived (Hilsenhoff 1991; Williams and Feltmate 1992). Most aquatic species are found in lake environments, but a few inhabit streams and rivers. Coleopterans are both primary consumers and predators. Because their habitats are also attractive to fish and birds, one usually does not find beetles in high densities.

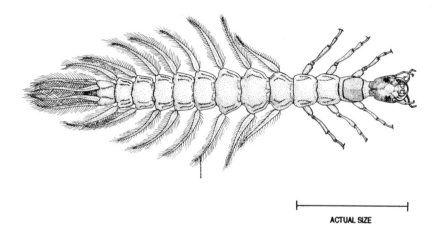

ACTUAL SIZE

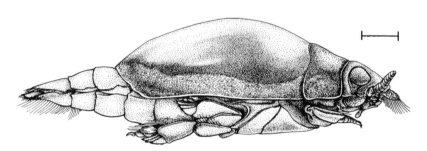

Figure 6.22. Beetle larva and adult.

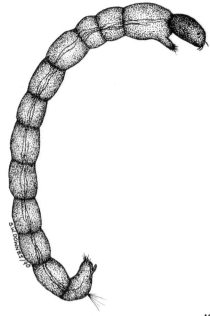

Figure 6.23. Midge larva.
(Drawing by S.W. Downes.)

ACTUAL SIZE

Flies (Diptera)

Flies (Order Diptera) are characterized by only one pair of functional membranous wings in the adult stage. Found in virtually every aquatic environment, they are often the only insects to be found in environmentally harsh freshwater habitats. Although the vast majority of species are terrestrial, those with aquatic immatures may be the predominant insects in many freshwater habitats. Over half of all known species of aquatic insects are dipterans; the chironomids (midges) constitute the largest family of freshwater insects. At least 30 families have aquatic or semiaquatic representatives, including some of the best known insect forms; mosquitoes, black flies, midges or gnats, crane flies and horse flies.

Virtually all aquatic insects that are of major economic importance or which serve as vectors of human disease are dipterans. Except for the economically important species, aquatic dipterans have been relatively well studied only in certain parts of Europe and North America, and very little is known about most tropical species. In aquatic species, only the larvae and pupal stages live in the water, the adults, with very few exceptions, are terrestrial. Larval body form is diverse, ranging from the primitive, cylindrical body with complete head capsule, to the maggot-like body with no externally discernible head. All types of feeding habits are exhibited by larval dipterans, sometimes within a single family. Their value as processors of organic matter is especially evident in aquatic systems receiving organic wastes, where "blood worms" may attain densities of many thousands of individuals per square meter. Aquatic dipterans have been widely used as pollution indicators. (Adapted from Williams and Feltmate 1992; Ward 1992).

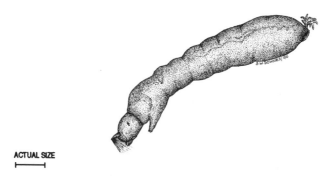

ACTUAL SIZE

Figure 6.24. Black fly larva. (Drawing by S.W. Downes.)

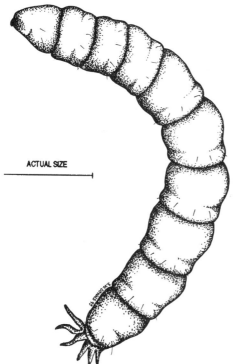

ACTUAL SIZE

Figure 6.25. Crane fly larva. (Drawing by S.W. Downes.)

Catchment Analysis Activities

Introduction

This chapter provides activities to guide learners in an analysis of their catchment. Section 1 takes a catchment-wide look at land use practices, while activities in Sections 2–5 focus on river reaches and sampling points. The activities relate to one another, building knowledge step by step; however, the activities are also designed to stand alone. Additional activities for physical and chemical water quality monitoring and analysis are included in Chapter 8 and in Appendix A.

> ➤ Section 1: Satellite, Mapping, and Land Use Activities (satellite images, aerial photos, mapping a catchment, rivers and people)
> ➤ Section 2: Riparian Activities (stream/river walk, verge and bank vegetation analysis, and bank erosion and stability)
> ➤ Section 3: Open Water, Pools, and Bends Activities (physical characteristics, land uses, water odors, and habitat assessment)
> ➤ Section 4: Phytoplankton and Macrophyte Activities (surveys and analysis)
> ➤ Section 5: Benthic Macroinvertebrate Activities (indices of water quality)

Each activity includes the following elements: objectives, materials, time, background information, procedures, conclusions, and follow-up questions. Use the descriptions to decide which activities to perform, keeping in mind what you hope to accomplish, the skills and needs of the group, how you are going to use the information, and your time-line. Each activity is labeled as optional or recommended. This status was determined by cost, significance, and difficulty.

Many of the activities include data sheets, which are found in Appendix B. These sheets provide an easy, organized manner of collecting data and are meant to be copied and used in the field. A list of the activities described in Chapters 7, 8 and 9 is shown in Figure 7.1.

Some of the activities include a quantitative rating system. This system provides a uniform method of rating various parameters and can be used as a tool to compare catchments. The quantitative rating includes four levels; you simply choose the number that best describes your site. The numbers correspond to the following ratings: 4 = *excellent*, 3 = *good*, 2 = *fair*, and 1 = *poor*. The quantitative rating system has been adapted from the excellent scheme developed by the Waterwatch Queensland Technical Manual (1994).

Many activities throughout this chapter require averaging multiple samples. To get the average of your samples, add the individual samples and get a total, then divide this total by the number of samples taken.

As always, safety is the first priority. Before conducting any of the activities, please see Chapter 5 for safety guidelines.

Catchment Activities:
Satellite, Mapping and Land Use

Overview

A catchment is an area of land that "catches" precipitation and over-land surface flow and sends it to a common channel. It is useful to begin a water quality monitoring project by studying the interactions of land and water in a catchment as a whole, before examining parts of the system. To better understand how people impact water quality and quantity, it is necessary to measure and observe the physical characteristics of the catchment and of its river. By taking a bird's eye view of the catchment through topographic maps and satellite images, it is possible to grasp the physical complexities of a catchment.

Remote sensing is the acquisition of information about objects by way of sensory devices which are remotely situated, such as in airplanes and satellites. Remote sensing has proven to be a powerful and valuable tool for analyzing land-use practices within a catchment. Aerial photographs and satellite images are useful in the detection and analysis of water pollution. The interface between land management and river systems, as well as information on droughts, floods, deforestation, agricultural practices, and land-river buffers is increasingly monitored through remote sensing.

The two major forms of remotely sensed data are aerial photos and satellite images. Both have benefits and drawbacks, and the preferred system depends on what information is needed. There are some universal strengths and weaknesses associated with both methods of remote sensing. Remote sensing allows large scale monitoring of the environment. A single image can cover a small catchment, allowing the user to see the "big picture" and to monitor a large area instead of small individual areas. Of

INDEX TO MONITORING ACTIVITIES IN CHAPTERS 7, 8 AND 9

Figure 7.1. Low-cost water quality monitoring activity index.

Figure 7.2. In studying any river system, it is critical to understand the physical characteristics of the entire catchment (photo by Martha Monroe).

course, the larger the area imaged, the lower the resolution, or detail, will be. The spatial resolution of satellite images is measured in units called pixels, which are the smallest pieces of information that can be detected. The resolution of some common satellites are as follows: NOAA Weather Satellite, 1 km x 1 km; Landsat Satellite, 80m x 80m; and the more recent French Spot Satellite, 10m x 10m.

Another major advantage of remote sensing is that it can detect a much wider range of the electromagnetic spectrum than the human eye can see. Various sensory devices can detect information about objects in the ultraviolet, visible, and infrared wavelengths. For example, the thermal band would detect the heat discharged from a heating plant. Because

Figure 7.3. Aerial photograph of channelized section of the Rouge River, Michigan, USA. Note a Combined Sewer Overflow (CSO) entering the river.

each satellite that is launched is used for many years, systems are designed to record over a range of spectral bands that will meet the needs of many users. One limitation of remote sensing is that it is difficult to make accurate measurements of small scale objects. Therefore, assessment programs should not rely on remote sensing alone. Remotely sensed images should always be "ground truthed," that is, followed by field observations and measurements to verify the data and interpretations.

Local watershed councils, planning bodies, universities, and state water agencies often have aerial photographs (Figure 7.3) or satellite images (Figures 7.4, 7.6 and 7.7) that are available for class use. For further information on obtaining aerial photographs or satellite images, please see Appendix A.

Activities 7.1–7.3 focus on the implications of changing land-use practices on the quality of a river system. Activities 7.4 and 7.5 seek to answer the questions: How large is the catchment? Where are the boundaries of the catchment? Where does the water that you use come from? How do you and the people who live in your community use nearby rivers or streams? What is the water quality in the area where you live? Was it always that way?

Figure 7.4. Near the bottom of this satellite image you can see several branches of the Rouge River forming a main branch prior to entering the Detroit River.

Activities for this section are listed below. Each activity has a label (recommended or optional) to indicate the author's judgement as to the universality of the activity (recommended) or potential difficulty in obtaining material (optional).

Activity 7.1 Your River from Beginning to End (Optional)

Activity 7.2 Changing Land Uses Over Time (Optional)

Activity 7.3 A Changing World (Optional)

Activity 7.4 Mapping a Catchment (Recommended)

Activity 7.5 Rivers and People (Recommended)

Activity

(Optional)

Your River from Beginning to End

Objectives

- Learn to interpret aerial photographs and satellite imagery.
- Develop a basic understanding of the structure of rivers and catchments.
- View the river holistically.
- Gain comprehension of catchments, headwaters, tributaries, drainage, and discharge.

Materials:

Slides, slide projector and/or copies of satellite images or aerial photographs, catchment maps, topographic maps of the watershed (if available), markers or crayons, and pencils; no Activity 7.1 data sheet in Appendix B.

Time:

Approximately 45 minutes.

Background Information

This activity is designed to provide students with a foundation of knowledge of rivers and water quality. It is important to understand the basic structure of rivers and catchments to fully understand the processes and factors that control river systems. Aerial photographs and satellite imagery (Figures 7.4 and 7.5) offer an opportunity to view the river as a whole. Through looking at images of the whole river, students will gain a better understanding of catchments, headwaters, tributaries, drainage, and discharge.

Procedures

1. Use a slide or provide multiple copies of an aerial photo/ satellite image of your river so that everyone can follow the

activity. Also, hand out catchment maps so that students can write in characteristics as the activity progresses.

2. A river begins at its headwater and flows downstream to its mouth. The headwater is the river's source and is located the farthest upstream. The mouth is where the river ends. Most rivers will end when they flow into another water body, such as a lake, ocean or another river. Locate the headwater and mouth of your selected river on the aerial satellite image of your catchment.

3. Look at the river channel. How long is the river? Is it straight or winding? Are there any distinctive features?

4. One important factor that will affect river water quality is the river's source. The source of a river can be underground water coming to the surface as springs, rainfall or snow melt, wetland drainage, glacial melt water, or the outlet of a pond or lake. Can you determine the source of your river by looking at the headwaters?

5. Locate your community on the map. Where does your community obtain its water? Locate the point on your map.

6. What kind of land uses along the riverbanks can be identified from the photo? How does the land change as you move down-river? How might this affect the river? Look especially at vegetation, development, and urbanization.

7. The flow of a river is important to the organisms that live within it. Although we cannot measure rates of flow from images, sometimes we can determine the calmness or turbulence of the flow.

   a. Aerial photographs are best for this, but satellite images can also be used. Try to detect differences in the flow of water in your image. Calm waters tend to look flat and uniform, while very rough waters will look bumpy or rough.

   b. If your image does not offer great enough resolution to determine roughness, you may still get a sense of the flow by looking at the river as a whole. Can you see the bottom of the river? Are there objects, such as rocks, protruding from the water that would result in turbulence? Does the width of the river vary considerably? If the river is wide and then narrows, the same volume of water will have to get through a smaller area. This would result in a more rapid, turbulent flow.

Discussion Questions

1. Discuss the effects on the whole river of drastic land use changes along the river bank at the headwaters, halfway down the river, and at the mouth. Which would be more detrimental?

2. Discuss the effects of land use practices within the catchment on people, water quality, river organisms, and other aspects of the ecosystem.

Activity

(Optional)

Changing Land Uses Over Time

Objectives:

- Study different land uses in the catchment.
- Develop an understanding of how land uses affect water quality.
- Gain awareness of human impact on water quality.

Materials:

Copies of an aerial photograph or satellite image include your catchment, transparencies, photographic tape, pens; no Activity 7.2 Data Sheet in Appendix B.

Time:

Approximately 30 minutes.

Background Information

Land use is a very important factor that influences water quality. Whether land is residential, agricultural, or industrial, land use decisions will be reflected in the water quality. Different land uses contribute to different types of water quality problems, such as farmland adding nutrients and pesticides to a river system; a highway bridge resulting in the runoff of oil, gasoline, or salt into a river; a residential area with septic tanks increasing biochemical oxygen demand; or factories discharging warm water directly into the river.

Pollutants also come from the atmosphere in such forms as particles, heavy metals, and acid rain. These atmospheric pollutants are due to human activity, and influence multiple catchments and bodies of water since air moves freely between catchments. When you look at the different land uses in your catchment, consider how they could add to atmospheric pollution which may pollute rivers miles down wind. Because land use is so important, it is essential that you know the land uses in your catchment and the impacts they are having on your water.

Note: Some of the data obtained through this activity can be applied to Activity 7.6 and Activity 7.10. It is helpful to have data sheets from Activity 7.6 and 7.10 (Appendix B) on hand.

Procedures

1. Work in pairs or larger groups, depending on the availability of images.

2. Place a transparency over the part of the image that includes your catchment, and tape it down. (If you need to move the transparency, carefully remove the tape starting with the end on the transparency to avoid tearing the picture.)

Figure 7.5. Three students noting land uses in the upper, middle, and lower parts of their catchment area.

3. Once you have the transparency properly placed, outline your catchment. The catchment's basic shape surrounds the tributaries or branching streams. The catchment's simplified boundary is an outline of the entire river system. (Note: When determining the true catchment one must account for the slope and elevation of the land; please refer to Activity 7.4.)

4. Starting at your river's headwaters and working toward its mouth, try to determine each land use. Mark them on the transparency.

Discussion Questions

1. What do you expect the water quality to be in the river in and around your community due to existing land use practices?

2. How have land use practices changed in your catchment over the past 10 years?

3. What has been the effect of these land use practices on the quality of your river?

Activity

(Optional)

Catchment Changes and River Trends

Objectives:

- Develop awareness of changes that have taken place in the catchment.
- Analyze water quality trends.
- Think about possible causes, impacts, and ways to prevent changes.

Materials:

Images of catchment over past years (at least 10 years ago to present, if possible), 1/4 inch (0.6 cm) grid (transparent), transparencies, colored pens, paper; no Activity 7.3 data sheets in Appendix B.

Time:

Approximately 60 minutes.

Background Information

Our environment is constantly changing due to naturally-occurring and human-caused factors. Some of these changes will only be temporary, others will be so minor that they will go unnoticed, while others will be so drastic that their impact will cause permanent effects. Although these changes are not always for the worse and have varying effects on different species, an awareness of them is essential for a fuller understanding of our closely integrated environment.

Photographs of the same area often are taken every few years, creating an extended record of land use changes over time. By comparing aerial photographs taken at various dates, one can detect human factors such as former landfills, dump sites, lagoons, pits, and above-ground storage tanks that currently are obscured by development or vegetation. In addition to identification of specific contaminant sources, analysis of land use changes over time can provide insights into general trends in water quantity and quality.

Procedures

1. Work in small groups with images of your catchment over past years.
2. How long ago was your earliest image taken? Record the time-span your images cover.
3. Spend some time becoming familiar with the images. Can you find your home or school in any, or all, of the images, or were they not built until later years?
4. What other common features are in your catchment? Are there buildings or other features that are evident in earlier images but not found in later ones?
5. The introduction of built features, such as housing developments, commercial industries and factories have the potential to affect water quality (Figure 7.7). Compare images consecutively, noting any new developments or expansion of older developments. Note the disappearance and replacement of any features.
6. The relative distance from the river of some features might influence their impact on the river water quality. Estimate and record the distances.
7. Define impervious and pervious land:
   a. Take your earliest image and tape a transparency to it. (If you need to move the transparency, pull the tape

Figure 7.6. Satellite image of the St. Clair River separating the United States from Canada prior to passing through the Walpole Island Reservation marsh. Note the river on the right side of the image carrying a heavy silt load.

off from the transparency first so that you will not tear the image.)

b. Use different colored pens or some other marking system on the transparency, to differentiate between the impervious and the pervious land.

c. Overlay the grid on your transparency.

d. Count the total number of grid squares that fall within the area of interest surrounding your catchment. For those squares that are not totally in the area of interest estimate to within one quarter of a square. Let this number equal t.

e. Estimate the number of squares that are impervious, again making all measurements to within one quarter of a square. Let this number equal n.

f. Calculate the proportion of impervious surfaces to total area by dividing n by t (n/t).

g. Repeat steps e and f for the pervious surfaces.

h. Now that you have calculated the proportion of impervious and pervious land for the first image, repeat this process for all images. If you can, do all the work on the same transparency so that you can visualize the change. Was there a significant change? What changes in water quality would you expect to result?

Discussion Questions

1. Discuss the possible effects of the changes you observed in your images on the water quality in your catchment.

2. Would any of the changes result in improved water quality?

3. What changes might you see if you had images from over 100 years ago?

4. How do you think the percentage of impervious to pervious surfaces affects water quality?

Activity

(Recommended)

Mapping a Catchment

Objectives:

- Understand the concept of a catchment.
- Map the boundaries of a local catchment.
- Show the locations of the channels that form your river system.
- Locate significant natural features within the catchment.
- Illustrate major land uses within the catchment.

Materials:

Geographic Survey maps, road maps, topographic maps, transparencies, pencils, crayons or colored pencils, paper (preferably for tracing), Activity 7.4 data sheets from Appendix B. (The map should include the entire catchment, or sub-catchment area, if your catchment is extremely large and difficult to encompass on one map. Helpful to have on hand for students copies of Activity 7.3 data sheets in Appendix B.)

Time:

Approximately 80 minutes.

Background Information

The first step in your assessment of water quality is to define the boundaries of your catchment. River quality is directly linked to the quality of the land that the river flows through. To understand the relationship between river quality and land-use practices, it is important to understand the concept of a catchment. A catchment is the total land area that contributes runoff (rainfall and snow melt) to a particular stream or river. It is the hydrologic unit that affects all living and non-living things within its boundaries. Catchments vary in size, from very small areas that drain water into small streams to huge catchments that collect water for large river systems. Within a large catchment (like the Great Lakes Basin or the Nile River Basin), each tributary is part of a smaller catchment. A series of catchments forms a drainage basin.

Reading a Topographic Map– Basic Terms and Conventions

1. Contour lines are generally brown; water features are blue; vegetation is green; cleared areas (fields, developed areas, and farmland) are white; and roads, buildings, and other non-natural features are black. Urban areas are gray.

2. All points along any one contour line are at the same elevation. Contour lines never cross each other.

3. Elevation, in meters above sea level, is indicated at intervals on contour lines and on the summit of many hills and mountains.

4. The difference in elevation between two adjacent contours is called the contour interval. It is usually given in the map legend. If the contour interval is 20 meters, the land climbs or descends 20 meters in elevation between a point on one contour and a point on the next.

Reading a Topographic Map– Recognizing Features on the Map

1. Slopes: Contour lines that are closely spaced represent steep slopes, and those that are widely spaced represent shallow or flat areas.

2. Hills: Hills and mountains appear as a series of successively smaller, irregularly shaped concentric circles. The smallest circle represents the highest point.

3. Water Flow: Water flow is perpendicular to contour lines. Streams tend to form in the V-shaped contours on side slopes, with the V's pointed in the direction of higher ground (i.e., upstream).

Procedures

1. Locate your local stream or river on a map (preferably a topographic map).

2. Select a spot on the map, as far downstream as possible, for a starting point. Next, locate the upstream ends of all channels that flow into your river above that point.

3. Draw a line that includes all of the branches or tributaries of your stream or river. If using a topographic map, the catchment boundary can be determined by connecting the high points and ridges surrounding all the branches or tributaries. This is your catchment. Be sure to distinguish it from other catchments.

4. Calculate the area of your catchment from the above map: Place a grid transparency over the catchment area. Count the number of boxes included with the catchment boundary. Determine measurement of distance on the map by using the map and its scale. To find the area multiply the centimeter scale by the number of boxes (#1 on Activity 7.4 data sheet in Appendix B).

5. Record the number of major branches within the catchment (#2 on data sheet).

6. Record the watercourse name (#3 on data sheet).

7. Record the types of watercourses within the catchment: Ephemeral (lasting a very short time), intermittent (not continuous), or perennial (continuous) (#4 on data sheet).

Figure 7.7. A satellite image of the channelized part of the Rouge River passing through the city of Detroit. The water treatment plant is located near the bottom of the image.

8. Average Watercourse Gradient (optional): Gradient or slope is a measurement of how much a watercourse descends in relation to how far it moves horizontally. In other words, slope is a way of quantifying how steep a river is. Topographic maps can aid in computing the average slope along the watercourse. Choose two points along the river. Point A is downstream or down slope, and point B is upstream or up slope (# 5 on data sheet).

a. A = the elevation at point A, B = the elevation at point B, and C = the distance between these two points

b. To calculate slope use the equation:

$$\frac{B - A}{C} = Slope \qquad \left(= \frac{rise}{run} \right)$$

c. Compute the slope for every 50 meters and then calculate and record the average.

9. Calculated Discharge (optional): Discharge is a measure of the volume of water passing a certain point over a specific period of time. The method for determining discharge shown here is called the "Embody Float Method." (# 6 on data sheet). The formula is as follows:

$$D = \frac{WZEL}{T}$$

D = discharge
W = average width of stream
Z = average depth
L = length of stream measured
T = time for float to travel length T
E = a constant:
  (0.9 for sandy/muddy bottoms)
  (0.8 for gravel/rock bottoms)

Note: The values for W, Z, L, and T may be taken from Activity 7.9. Therefore, discharge may be calculated after the completion of Activity 7.9.

10. Use your own and other people's knowledge of your region to identify on the map the major land uses within the catchment. (General Development Maps and Zoning Maps, available from county and township governments in the U.S., are helpful in identifying land use.) (#7 a–h on data sheet)

11. Record on the data sheet the area percentage of the land uses (Activity 7.9 may be helpful in calculating these percentages). Major land uses may include: agriculture, urban/suburban, industrial, mining, logging, grassland, and forested (#7 a–h on data sheet).

12. Locate or draw on the map significant natural features (including forests, grasslands, deserts, mountains, valleys, plateaus, ridges, lakes, marshes, deltas, etc.).

13. The finished map will be used to aid in further activities.

Discussion Questions

1. Where does the water in your catchment come from (rain and snow, ground water, glacial meltwater, from wetlands, other sources)? Are the streams and rivers in the watershed present all year, or do they dry up during hot seasons?

2. What are some of the major land uses in your catchment? How might these different land uses affect the river?

3. What are the percentages of land use types in the catchment? Do land use changes occur as you move from the headwaters to the mouth of the river?

4. How would human activities in one part of the catchment, for example dumping of sewage or industrial wastes, affect other parts of the catchment?

5. What is the average slope of your river reach? How does this slope relate to the velocity of the river?

Activity

(Recommended)

Rivers and People

Objectives:

- Develop interviewing skills.
- Design a questionnaire.
- Compile the results of interviews and piece together the history of the river.

Materials:

Pencil, paper, tape recorder (optional); no Activity 7.5 data sheet in Appendix B.

Time:

Approximately 80 minutes class time; 40 minutes to design questionnaire and 40 minutes to discuss results. (Students should do the actual interviews outside of class.)

Background Information

Rivers and streams are the focal points of human activity around the world. People who live close to the natural world come to know local watercourses intimately. Within every catchment and along every river, there are people who have watched rivers flow for most of their lives. These people have knowledge about the community and river that is just waiting to be tapped.

In this activity, students will interview such people to learn more about the river. History comes alive through interviews. Discussions with real people about their impressions and experiences with the river is often much more meaningful than simply reading a book or article.

It will be useful to have the data sheets from Activity 7.10 on hand when performing the interviews.

Procedures

1. Have students work alone or in groups to develop a series of questions they would like to ask someone about the river or stream. Below are some sample questions:

   - *How long have you lived in this area?*
   - *What do you remember the river being like, when you were my age?*
   - *Did you use the river in different ways than we use it now?*
   - *What are your hopes for the river in the next century?*

2. Have students interview several people about the river (they can do this as a homework assignment over a weekend, or perhaps over the course of an entire week). The following suggestions will help students to conduct an effective interview.

   a. Try to interview older people, who may have a lot of knowledge about local history. Family members and neighbors may be a good place to start in recruiting people to interview.

   b. Inform the person you are planning to interview of the purpose of the interview, how long it will last, and how the interview results will be used.

   c. Give them an idea of the types of questions you will be asking.

   d. Try to draw out specific examples from people who make general statements. For example, if someone were to tell you that they used to play in the river, ask them: What did you do in the river? Did you swim, fish, collect bugs?

   e. Role-play interviewing other students before conducting the real interviews.

Discussion Questions

1. Discuss your interview findings as a whole group. How do the students' views of the river differ from those of the people whom they interviewed?

2. In what ways is the river of today like the river of yesterday (10, 20, 30, 40 years ago)? How is it different?

3. Were there any common experiences with the river? How did experiences differ?

4. How might some of the changes come about?

5. How did it feel to learn about the river from other people?

Riparian Activities

Overview

In section 1, the entire catchment was the focus. In this section you will have the opportunity to explore a much smaller portion of the catchment: a longitudinal segment of a local stream or river called a reach. The riparian environment within the reach is the area of focus for this section. The riparian environment is made up of the banks of the watercourse and the verge (the strip of land extending 20–30 meters from the bank). Plants that grow on the sides of watercourses are extremely important for the health of the catchment environment. Riparian environments function within the catchment in four significant ways. They contribute to the ecology of both the aquatic environment and the surrounding catchment environment, to soil stability, to water quality by serving as a buffer or filter for catchment runoff and nutrient uptake.

Activities for this section include:

Activity 7.6 Stream/River Walk (Recommended)
Activity 7.7 Bank and Verge Vegetation Evaluation (Recommended)
Activity 7.8 Bank Erosion/Stability Evaluation (Recommended)

Selecting a River Reach

The first step in performing any of the activities in this section is to designate the stream or river reach (or reaches) for student groups to walk and survey. Visit potential sites before exploring with the students. Use the following criteria when selecting reaches:

➤ Accessibility and safety are two important criteria to consider when selecting reaches. Be certain to obtain permission if the river access is on or through private land.

➤ Each reach should be bounded by conspicuous reference points such as bridges, monuments, or distinctive natural features, so that students know where the reaches they will investigate begin and end. Consulting a detailed local map and physically walking the area are helpful at this stage.

➤ Half a kilometer (0.3 miles) is a good length for each reach, but you should modify lengths based upon the conditions at your site.

➤ Stream or river reaches should be coded or labeled. This will help you to organize the results from each student group. The labeled segments could include the first two letters of the river or stream name, e.g. "RO" for Rouge River. This would be followed by a number that indicates relative position along the stream or river. The segment at the mouth of the Rouge River, USA, might be labeled "RO1," the next segment upriver would be labeled "RO2," and so on.

Activity

7.6

(Recommended)

Stream/River Walk

Objectives:

- Engage students in a visual survey of a river or stream that involves mapping and recording information.
- Stimulate students to see the river or stream in new ways.
- Uncover potential sources of pollution that should be studied further.
- Learn methods of communicating locations and points on maps to others.

Materials:

Clipboards, pencils, map and data developed in Activity 7.4, (Optional: zoning, land ownership, and historical maps and aerial photographs of the catchment); no Activity 7.6 data sheet in Appendix B.

Time:

Approximately 120 minutes total; 40 minutes each to complete the narrative, stream walk, and classroom activities.

Background Information

On your stream or river walk you will investigate specific details about the water quality and local land uses of your reach, including: color and appearance of the water; signs of fish and other organisms; potential pollution sources, such as discharge pipes, dumps, or construction sites; and the nature of the stream or river bottom.

It may be difficult in some areas to get to the river bank, or to walk along the river. The river may be too deep or wide to see the river bottom, or the opposite side. However, we encourage you to explore and investigate as best you can.

This activity includes three components that are meant to complement one another. Activities 7.6, 7.7 and 7.8: a narrative account by the students of their walk, a stream walk map, and classroom activities.

Procedures

1. *Narrative (a written account of the stream or river walk):*

   Richness of detail about land uses bordering the watercourse, about colors and odors, about human usage of the river, about suspected pollution sources, and about students' perceptions of being at the river emerge from narratives. The narrative is most effective when students use descriptive language that gives color and texture to their perceptions.

   a. Have students explore their reach, noting their thoughts, feelings and impressions in journals or notebooks on-site.

   b. Encourage creativity and emphasize feelings and perceptions about the river. You will have ample opportunity to be more analytical in other parts of this activity.

2. *Stream Walk Map:*

   a. As a whole group, create a system of symbols that denote land uses and built structures.

   b. Each group can be given its own hand-drawn, photocopied or mimeographed copy of a detailed base map, showing prominent landmarks. Make sure the scale of the map is such that each reach is shown in sufficient detail (if necessary, enlarge the map). Each student group should generate simple maps of their reaches by walking along the banks and noting natural and built features.

3. *Classroom Activities:*

   a. Once all groups have collected their data, have them add their findings to a large scale catchment map. Use colored pencils or crayons to illustrate major land uses and other results from the field work.

   b. Have each group report to the entire class what they found in their reach.

   c. Share passages from student narratives.

Figure 7.8. Hawaiian students recording bank erosion and instability on the Maunawili River in Hawaii.

Discussion Questions

1. What do you think about the overall quality of your river? How do you feel about its current state?

2. Based upon your field work, or things you learned elsewhere, what are some potential problems facing your river today?

3. Did you discover any potential sources of pollution?

4. Is there any evidence of human activities which have altered flow or course of the river (channelization, dams, other)?

Activity

(Recommended)

Bank and Verge Vegetation Evaluation

Objectives:

- Understand the importance of the vegetation of the riparian zone.
- Determine the quality of the bank and verge vegetation.

Materials:

Local map, writing utensils, Activity 7.7 data sheets in App. B.

Time:

Approximately 30 minutes.

Background Information

The riparian zone is the area directly influenced by a body of water, such as along streams and lakes. It has a characteristic vegetation that differs from upland areas.

The vegetation located in the riparian zone affects the health of the catchment environment in a variety of ways. Riparian vegetation provides food for animals and other organisms living in and near the water. This food mainly comes in the form of falling leaves, branches, and logs. Insects attracted to the riparian zone become part of the food supply when they fall into the water or lay their eggs there. Many animals depend upon the logs and branches that fall as a source of shelter. These larger forms of debris also affect the flow of the water and thus provide better living conditions for some of the animals. Overhanging trees provide shade that lowers water temperatures, and helps prevent an imbalance in the amount of plant growth (*Streamwatch Manual*, Appendix D). The riparian vegetation also protects the bank from erosion. Plants have the ability to soak up nutrients and pollutants. Thus they act as a buffer to reduce the amount of pollutants entering the water.

The riparian zone is sometimes divided into two parts: the *bank* area extending from the water's edge to the top of the streambank, and the *verge* zone from the top of the streambank to the upland limit of the riparian zone (see Figure 7.9).

Procedures

Record observations on Activity 7.7 data sheet in Appendix B.

1. Bank Vegetation: Identify the density, and composition of the bank vegetation: barren, grasses, brush, deciduous, conifer, etc. Check all boxes on the data sheet that apply.

2. If known, list the names of all species of bank vegetation found growing in your reach, in the spaces provided on the data sheet. If the name is unknown, draw a picture to illustrate the species.

3. Verge Vegetation: Identify the clusters, density, and composition of the verge vegetation (to approximately 30 meters from the bank): barren, grasses, brush, deciduous, conifer, etc. Check all boxes on the data sheet that apply.

4. If known, list the names of all species of verge vegetation found growing in your reach, in the spaces provided on the data sheet. If the name is unknown, draw a picture to illustrate the species.

5. If you have access, do steps 1–4 on each side of the river. Concentrate on only one side of the river at a time.

6. Compare the vegetation you have identified to a field guide or list of expected native species for your area.

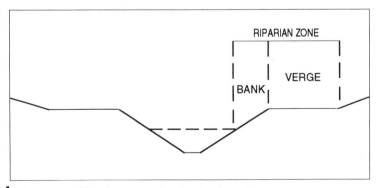

I **Figure 7.9.** Riparian zone showing bank and verge areas.

Figure 7.10. The Mary River in Australia carrying a heavy load of silt and indicating severe bank damage and instability following a storm.

Conclusions

Rate the vegetation growing on the bank and verge of your sample reach using the following scale. Select the number that most closely describes the vegetation you observed.

BANK VEGETATION QUANTITATIVE RATING	SCORE _____
4 (excellent)	Vegetation in undisturbed site
3 (good)	Vegetation, mildly disturbed
2 (fair)	Vegetation moderately disturbed
1 (poor)	Vegetation severely disturbed

VERGE VEGETATION QUANTITATIVE RATING	SCORE _____
4 (excellent)	Vegetation present and canopy intact
3 (good)	Vegetation and canopy virtually intact
2 (fair)	Vegetation clearly disturbed
1 (poor)	Cleared land or urban development

Discussion Questions

1. Did the kinds of vegetation indicate a disturbed site or reach?

2. Was there vegetation overhanging the river? Could you see any vegetation (leaves, twigs, fruits) in the water?

3. Did the density and composition of the vegetation differ for the different sides of the riverbank?

4. How did the vegetation change from riverbank to the verge?

Activity

Bank Erosion/Stability Evaluation

(Recommended)

Objectives:

- Recognize evidence of bank erosion (bare soil, movement).
- Understand the consequences of bank erosion.

Materials:

Pencils, Activity 7.8 data sheets in Appendix B.

Time:

Approximately 30 minutes.

Background Information

Riparian vegetation exerts major control over bank stability. One study found that bank sediment which was made up of 16–18 percent roots, with a 50 mm deep root mat on the bank surface, had 20,000 times more resistance to erosion than comparable bank sediment without vegetation (*Waterwatch Queensland Technical Manual*, 1994). Water quality and the aquatic environment improve where erosion is prevented. (For a better understanding of how increased sediment in the water impacts the aquatic ecosystem, please read Chapter 4 and see Activity 8.8, Turbidity.)

Figure 7.11. A student estimating the percent of bare soil in the riparian zone which is not bound by plants or their root structure.

People may contribute to warmer water by causing soil erosion along the river bank. Erosion can be caused by the removal of trees and other vegetation, poor farming practices (plowing near the streambank), highway construction, and other construction.

Soil erosion is an important factor in river temperature because erosion increases the amount of suspended solids carried by the river. A large amount of suspended solids turn the water turbid or very cloudy. This cloudy water absorbs the sun's rays, which warms the water.

Procedures

Record observations on the Activity 7.8 data sheets in Appendix B. For the following observations, combine both sides of the river.

1. Bare Soil: estimate the percentage of the area in the riparian zone which is bare soil not bound by plants and their root structures or covered in concrete or rocks (Figure 7.11). These bare areas could be due to people access, roads or crossings, clearing, soil erosion or undercutting. Record your estimate on the data sheet.

2. Bank Slope: note the steepness of the bank slope. Is it steep, moderate or slight? Check the appropriate box on the data sheet.

3. Bank Stability: estimate the amount of erosion that is present on the bank. Rate the bank erosion using the qualitative rating system in the conclusions/indications section below. Record the rating you have selected on the data sheet.

4. Slumping or Bank Movement: rate the slumping and movement of the bank using the quantitative rating system in the conclusions/indications section below. In some areas the removal of vegetation leads to slumping and movement of the banks. Record your selection on the data sheet.

Conclusions

Rate the bank soil, bank slumping and movement, and bank erosion of your sample reach using the following scale. Select the number that most closely describes the bank you observed.

PERCENT BARE SOIL		SCORE _____
4 (excellent)	0–10 percent	
3 (good)	11–40 percent	
2 (fair)	41–80 percent	
1 (poor)	81–100 percent	

EXTENT OF SLUMPING AND MOVEMENT		SCORE _____
4 (excellent)	no movement	
3 (good)	slight movement	
2 (fair)	moderate bank collapses	
1 (poor)	severe bank failure with extensive cracking and fall-ins	

AMOUNT OF BANK EROSION		SCORE _____
4 (excellent)	stable, no sign of any bank erosion	
3 (good)	very occasional and very local erosion	
2 (fair)	some erosion evident	
1 (poor)	extensive erosion	

Now average the ratings you have selected to give an overall rating for the bank erosion and stability.

Discussion Questions

1. Are there areas of bare soil along the river? What are the causes of bare soil?

2. Is there evidence of erosion of the banks?

3. Soil erosion can lead to many water quality problems. Among them is sedimentation, or the settling of suspended sediments to the river bottom. Another is increased turbidity. Is there an increase in turbidity downstream from the eroded banks? How might you find out?

Open Water, Pools, and Bends Activities

Overview

Open water, pools, and bends are some of the physical characteristics of flowing water that shape the nature of the aquatic plants and animals there, as well as the human uses of the river.

The velocity of the current and its discharge determine the kinds of substrates found, which in turn influence the benthic macroinvertebrate community. Human uses of rivers are closely tied to seasonal changes in flow and the hydrology of rivers.

Human-caused changes to the river channel include dredging, channelization, construction of dams and weirs, and drainage of adjoining wetlands (see Chapter 3 for a detailed discussion of these).

The activities for this section include:

Activity 7.9   Physical Characteristics (Recommended)
Activity 7.10   Primary Uses and Impairments (Recommended)
Activity 7.11   Water Odors and Appearance (Recommended)
Activity 7.12   Habitat Assessment (Recommended)

Activity

(Recommended)

Physical Characteristics

Objectives:

- measure the physical characteristics of the river.
- understand the relevance and impact of physical characteristics.

Materials:

Pencils, tape measure, string, several oranges (or similar floating objects), thermometer, waders, stop watch, Activity 7.9 data sheets in Appendix B.

Time:

Approximately 120 minutes.

Background Information

As a river flows from its headwaters to its mouth, the water temperature, water volume, bottom consistency, and quality of food vary considerably. Some of these changes are related to the type and use of the land the river flows through; some are a function of increasing stream order.

The current or flow of a river is a major factor in shaping the river ecosystem. Aquatic plants and animals depend upon a river's current to bring food and nutrients from upstream, and to flush wastes downstream. Measuring velocity and calculating discharge can give much information about a river system. Part of this activity will include measuring four factors that control the speed of a river: depth, the slope or steepness of the land, the width of the stream channel, and the roughness of the river bottom.

River discharge is also related to the climate of an area. During dry periods, flow may be severely reduced, and water temperatures will be warmer. The combination acts to reduce dissolved oxygen levels in water. Wet weather or melting snow increases flow and creates the possibility for greater mixing of atmospheric oxygen.

Observations made during this activity will help determine how the physical characteristics change along the river and the impacts they have on the river system.

Procedures

Divide into groups. Each group can perform a few of the tasks, then share their information.

1. Record on the data sheet the location of your individual observation site. It will be helpful to use the map created in Activity 7.4.

2. *Stream Type:* At your observation site, is the stream straight, meandering, braided, channelized, or pool/riffle? Check the appropriate box on the data sheet. (These terms are discussed in Chapter 4.)

3. *Today's weather:* Record today's weather conditions (clear, sunny, overcast, rain, showers, storm, etc.). Recent rainfall can affect flow, clarity, and amount of water in a stream.

   Some helpful definitions:

   Rain: up to 0.76 cm (1/3 inch) in 24 hours; light steady rainfall.

   Showers: 0.76 cm to 2.54 cm (1/3 inch to 1 inch) in 24 hours; intermittent and variable in intensity.

   Storm: 2.54 cm (1 inch) or more in 24 hours; usually accompanied by winds.

4. *Last Precipitation:* Record the last precipitation date, amount (cm), and duration (hours).

5. *Recent Weather:* Record recent weather that may have some affect on the water conditions, especially any major storms. Be sure to record the date of this weather condition. Recent newspapers may be helpful.

6. *Air Temperature:* Using a thermometer, measure the temperature of the air in degrees Centigrade. Do this five times in the same location, and record the average on your data sheet.

7. *Water Temperature:*

   a. Lower the thermometer four inches below the surface of the water.

   b. Keep the thermometer in the water until a constant reading is obtained (approximately two minutes).

Figure 7.12. Note the meandering course of the river and percent of surface plant cover along this wilderness stretch of river in Canada.

    c. Record your measurement in degrees Centigrade (see chart in Activity 8.5 Temperature, to convert Fahrenheit to Centigrade). Take five measurements at the same location and record the average on your data sheet.

8. *Average Stream Width:* Measure the stream's width from bank to bank at 5 locations along the observation site. Record each measurement on the data sheet. It is not necessary to compute the average on-site; it can be done later.

Note: If the stream is deep or fast moving, do not attempt to wade across to measure width. If there is a bridge nearby, you may be able to measure the width of the river across it. An estimate is fine. To estimate, lay a 1 meter length of string on the ground in front of you, perpendicular to the stream and estimate how many times the piece of string would fit across the river from bank to bank. Decide which width class your river would belong to and record this on the data sheet, indicating that it is an estimated measurement.

Width classes: <2m, 2–5m, 5.1–10m, or >10m.

9. *Average Stream Depth:* (For shallow rivers only) Wade into the river (wearing waders). Do not go out any deeper than thigh-deep. Using a ruler or tape measure, measure the

depth of the water in 5 locations and record your measurements on the data sheet. Again, it is not necessary to compute the average at this time.

Note: It is important to measure stream depth at the location of features such as riffles, runs, and pools (see Glossary). Label these features on your data sheets. This data can be used to complement the habitat assessment section. (To save time, average stream depth and channel slope measurements can be taken together.)

10. *Surface Velocity:*

   a. Use a tape measure along the stream or riverbank to mark a section at least 20 meters in length.

   b. Position someone at the upstream and someone at the downstream ends of the marked section.

   c. Release an orange into the main current at the upstream end of your marked length.

   d. Use a stop watch to time the passage of an orange from the beginning to the end of the marked length.

   e. The downstream person should yell when the orange floats by the end point to inform the time recorder.

   f. Repeat this test 3 to 5 times and average the results.

   g. Calculate the velocity in meters/second and record on the data sheet.

Note: An orange works well because it floats more or less in the zone of maximum velocity (just below the surface). However, a similar object may be used in its place. Do not take risks trying to retrieve your orange; it will not harm the river if it is swept downstream.

11. *Bankfull Width:* Most stream surveys are conducted during low discharge periods, so the width at the bankfull stage must be measured from an imaginary line that extends across the channel at the elevation of the flood plain. Some judgment will be required to predict the bankfull stage.

12. *Channel Slope:* The channel slope may be determined by measuring the vertical drop along a measured length of the stream bed at several locations at the site.

$$\frac{B - A}{CS} = C \text{ (Channel Slope)}$$

Wade into the water (wearing waders and a life preserver if possible). Measure the depth at point A. While standing at point A (upstream), hold the end of a tape measure and have a second person holding the other end of the tape measure go to point B (downstream). C is the distance between these two points. Measure the depth at point B. Subtract A from B and divide your answer by C. This will give you the channel slope (CS) for distance C. Do this procedure in 5 locations and compute the average.

13. *Channel Cross-Section:* Is your channel rectangular, U-shaped, V-shaped, or other? Please check the box on the data sheet which matches the shape of the stream channel. If you are unable to see the shape of the bottom and banks, please estimate. You can base your estimate on the flow of water. The slower the water in the middle of the stream, the flatter the bottom (*EPA Streamwalk Manual*, July 1994).

14. *Watercourse Bottom:* What is the predominant inorganic and organic substrate of your river?

Check the appropriate boxes on the data sheet.

Inorganic:

bedrock	
boulder	(> 25.5 cm or >10 inch diameter)
cobble	(6.5–25.5 cm or 2.5–10 inch dia.)
gravel	(2 mm–6.5 cm or 0.1–2.5 inch dia.)
sand	(0.005–0.20 cm or 0.002–0.079 inch dia.)
silt	(soft, fine sand)
clay	(fine sand with a sticky texture).

Organic:

muck–mud	(black, very fine)
pulpy peat	(unrecognizable plant parts)
fibrous peat	(partially decomposed plant material)
detritus	(sticks, wood, coarse plant material)
logs, limbs	(large pieces)

What percentage of the site substrate is inorganic material? What percentage is organic material? Record the percentages on the data sheet.

15. Frequency of flooding: Record the flooding patterns of your river. If unknown, give the best estimate: none, rare (10 to 20 years), occasional (5 to 10 years), frequent (1 to 5 years) or seasonal. Please check the appropriate box on the data sheet.

16. Watercourse channel alteration (if known): At your site, or up or downriver, has the channel been dredged or channelized? Is there a dam/weir or wetland drainage? Check the appropriate boxes on the data sheet, and if known, the dates any alterations occurred.

Discussion Questions

1. Watercourse bottom, channel cross-section, channel slope, and surface water velocity are interrelated. How are these observations and measurements related at your sampling site?

2. Could you predict, given the shape of the channel cross-section and the watercourse bottom, what the water velocity might be like?

3. Is there evidence of human-caused changes in the channel or watercourse bottom?

Activity

(Recommended)

Primary Uses and Impairments

Objectives:

- Study different land uses in the catchment.
- Develop an understanding of how land uses affect water quality.
- Gain awareness of human impact on water quality.

Materials:

Pencils, Activity 7.10 data sheets in Appendix B.

Time:

Approximately 30 minutes.

Background Information

Land use is a very important factor that influences water quality. Whether land is residential, agricultural, or industrial, the activities that take place will be reflected in the water quality. Different land uses contribute to different types of water quality problems. Pollutants can be introduced to a river system through point sources, such as pipes, or non-point sources, such as run off or atmospheric deposition in the form of either rain, snow, or dry deposition.

Although some of these pollutants can occur naturally, such as those from volcanic eruptions, most of them are the result of human activities. Very few people consider that the waste water from washing their car might end up in a river and kill fish; that a parking lot replacing an abandoned field means more runoff containing harmful debris; or that a farm's fertilizers might impact life in a river through the depletion of oxygen. It is essential to know the land uses in your catchment and the impacts they have on water quality. It is important to understand the roles that an individual, a family, a school, or a community can play in land use decisions (see Chapter 3).

Information for this activity may be obtained through personal observation or by asking local residents questions. Use the data sheets from this activity when performing the interviews for Activity 7.2.

Procedures

Record your observations on the data sheets for Activity 7.10 Primary Uses and Impairments (Appendix B).

1. What is the population served by the water in your catchment?

2. What are the primary uses by humans of this water? Examples include drinking water supply, bathing, recreation (swimming, fishing, etc.), washing clothes, agricultural water supply (irrigation, livestock, etc.), transportation (motorized boats, human-powered boats, shipping goods, etc.), industrial water supply, waste disposal. Check the appropriate boxes on the data sheet.

3. Are there any apparent water use impairments? If yes, please explain. Are the impairments caused by: agricultural runoff, housing, livestock yards, cropland, pasture, inadequate or overloaded wastewater treatment facilities (primary, secondary, tertiary, other), logging runoff, failing septic tanks, industrial discharge, mining runoff, golf courses, irrigation problems, construction, or other? Check the appropriate boxes on the data sheet.

Figure 7.13. Are there any land use practices in this photo that might have adverse effects on the quality of water?

Note: Some water use impairments may not be realized until after the physical-chemical tests in Chapter 8 have been performed. Return to this section after the performance of those tests to see if any further data may be added.

Discussion Questions

1. Can you identify the potential point source or non-point sources that cause impairment of use?

2. Which indicators should be measured to confirm the presence of pollutants?

Activity

(Recommended)

Water Odors and Appearance

Objectives:

- Determine water and soil odor.
- Determine water appearance.
- Understand the possible indications of odors and colors.

Materials:

One large-mouthed jar, one clear glass jar, white sheet of paper (or inside cover of manual), pencils, Activity 7.11 data sheets in App. B.

Time:

Approximately 30 minutes.

Background Information

Odor in water may be due to natural or human causes. Natural causes include decaying weeds and algae or the presence of micro-organisms. When organic matter decomposes, gases like ammonia and hydrogen sulfide are given off. Sewage and industrial wastes contain halogens, sulfides, or other chemical compounds and are also responsible for odor in water. Odor is undesirable in drinking water and certain industrial processes.

Color in river water may result from naturally occurring materials such as soil particles, dissolved or suspended clay, or decaying organic matter (tannins, peat, algae, fungi and weeds). Some industries, such as textile and leather processors, which use colors in their manufacturing, are also responsible for adding color to water bodies. Seepage from a wastewater treatment facility may give the water a green, green-blue, brown, or red appearance. Some of these color sources are harmful, some are not. Water appearance is sometimes a good indicator of local land use practices.

Soil acts as a filter as water passes through it. Thus, contaminants in the soil may end up in the water body. Soil smell will suggest pollutants that might exist only at trace levels in the water. The smell and appearance of drinking water and aquatic foods help determine the acceptability of a body of water for recreational uses, or warn of potential health hazards. Both can be invaluable, no cost means to assess water quality.

Procedures

1. *Water odor:*

    a. To assess the odor of the water, collect a sample of water in a large-mouthed jar.

    b. Use your hand to wave the air above the water sample toward you.

    c. Use the list of odors to describe what you smell. Record the type and intensity (faint, distinct, or strong) of the smell on the data sheet.

2. *Soil odor:*

    a. To measure the odor of the bank soil, disturb bank sediment and note any odor it gives off.

    b. Use the list of odors to describe what you smell. Record the type and intensity (faint, distinct, or strong) of the smell on the data sheet.

    Note: Do these procedures on the site immediately after collecting the sample. Samples may lose their odor over time. Students testing for odor should not have a cold or other sinus problems. Sense of smell varies from person to person. For good results, have several students smell the samples separately.

Helpful odor classifications include:

Chemical:	Chlorine Sulfur (hydrogen sulfide, rotten eggs)
Musty:	Decomposing Straw Moldy (damp cellar)
Harsh:	Fishy—Uroglernopsis, Dinobryon (dead algae) Septic—stale sewage
Earthy:	Peaty Grassy
Aromatic:	Spicy—camphor, cloves, lavender, lemon
Balsamic:	Flowery—geranium, violet, vanilla

(*Standard Methods for the Examination of Water and Waste Water,* 17th edition. 1989. American Public Health Association, Publications Office, 1015 15th St., NW, Washington, DC 20005).

3. *Water Appearance:*

Verbal descriptions of apparent color can be unreliable and subjective. If possible, use a system of color comparison that is reproducible. By using established color standards, people in different areas can compare their results. See Appendix B for information regarding two widely-used color systems.

To make color comparisons:

a. Take a sample of water in a clear glass jar.

b. Visually inspect the sample in adequate light, against the inside cover of this manual.

c. On the data sheet, check the box next to the color that best describes what you see. If using a color system or scale, match to a color standard and record on the data sheet the reference number of the color standard yielding the best match. Be sure to report the system of color standards used along with your observations.

Conclusions

The following guide to odors and appearance of water and soil is helpful in identifying potential pollutants or sources of pollution. Identifying pollution sources is the first step in further monitoring and in taking appropriate action.

Odors:

Sulfur (rotten egg):
May indicate the presence of organic pollution, such as domestic or industrial wastes.

Musky:
May indicate presence of sewage discharge, livestock waste, decaying algae, or decomposition of other organic pollution.

Harsh:
May indicate the presence of industrial or pesticide pollution.

Chlorine:
May indicate the presence of over-chlorinated effluent from a sewage treatment facility or a chemical industry.

No unusual smell:
Not necessarily an indicator of clean water. Many pesticides and herbicides from agricultural and forestry runoff are colorless and odorless, as are many chemicals discharged by industry.

Figure 7.14. What positive and negative land use practices along this sector of the Rhine Valley in Germany could affect water quality?

Water Appearance:

Green, Green-Blue, Brown or Red:
Indicates the growth of algae, which is usually caused by high levels of nutrient pollution. Nutrient pollution can come from organic wastes, fertilizers, or untreated sewage.

Light to Dark Brown:
Indicates elevated levels of suspended sediments, giving the water a muddy or cloudy appearance. Erosion is the most common source of high levels of suspended solids in water. Land uses which cause soil erosion include mining, farming, construction, and unpaved roads.

Dark Reds, Purple, Blues, Blacks:
May indicate organic dye pollution from clothing manufacturers or textile mills.

Orange-Red:
May indicate the presence of copper, which can be both a pollutant and naturally occurring. Unnatural occurrences can result by acrid mine drainage or oil-well runoff.

Blue:
May indicate the presence of copper, which can cause skin irritations and death of fish. Copper is sometimes used as a pesticide, in which case an acrid (sharp) odor might also be present.

Foam:
Excessive foam is usually the result of soap and detergent pollution. Moderate levels of foam can also result from decaying algae, which indicates nutrient pollution.

Multi-Colored (oily sheen):
Indicates the presence of oil or gasoline floating on the surface of the water. Oil and gasoline can cause poisoning, internal burning of the gastrointestinal tract, and stomach ulcers. Oil and gasoline pollution can be caused by oil drilling and mining practices, leakages in fuel lines and underground storage tanks, automotive junk yards, nearby service stations, wastes from ships, or runoff from impervious roads and parking lot surfaces.

No unusual color:
Not necessarily an indicator of clean water. Many pesticides, herbicides, chemicals, and other pollutants are colorless or produce no visible signs of contamination.

Discussion Questions

1. Based on your observations, what conclusions can you draw about the quality of your water samples?
2. What kind of pollutants might be present?
3. From the information you collected in other activities, what land uses might contribute to the odor and appearance you observed?

Activity

(Recommended)

Habitat Assessment

Objectives:

- Evaluate the variety of habitats at the site.
- Understand the importance of pools, riffles, bends, snags and undercut banks.

Materials:

Pencils, Activity 7.12 data sheets in Appendix B.

Time:

  Approximately 40 minutes.

Background Information

This activity teaches students to evaluate the variety of habitats available to organisms at your site. A stream with pools, bends, rocks, undercut banks, and snags (fallen branches, small log piles) contains better habitat for development of a diverse aquatic community than a straight or uniformly deep stream. These features are important for providing food, shelter from predators, and breeding places for animals in the water. They enable plants and animals to remain stationary, which otherwise might be swept downstream. In addition, protruding logs and snags provide roosting sites for birds and semi-aquatic wildlife.

Riffles and rapids create water turbulence, increasing aeration. Higher levels of dissolved oxygen should be found in and around riffles and rapids. Wetlands, weed beds, and pools often act as breeding grounds, critical for repopulation of the biota. Wetlands also act as natural filters, capturing pollutants before they can be introduced into the watercourse.

Overhanging trees and shrubs provide food for the animals in the water when they drop their leaves, fruit, and flowers, and when insects and spiders fall from them into the water. These plants also control water temperature by providing shade.

Procedures

Record data on the Activity 7.12 data sheets (Appendix B).

1. Select a site or reach of river that overlaps the stream or river walk site observed in Activity 7.6. Notice the places where plants and animals could live: pools, riffles, rocks, undercut banks, log piles, human-made objects (pilings, bridges), wetlands, and weed beds. Check all boxes that apply on the data sheet.

2. Animals are an important part of a catchment ecosystem. List the names of all the fish, reptiles, and birds you see during the time you are at the site. If you cannot identify a species, draw a picture to illustrate it.

3. Cross-Sectional Drawing: In the box provided on the data sheet, draw a cross-section of the water under investigation. Draw and label the locations of:
   • bank/flood plain on both sides of the river
   • shape/steepness of land, above water level
   • shape of channel, below water level
   • any eroded areas of river bank/flood plain

- types of vegetation
- visible substrate materials: soil, sand, gravel and boulders
- human-made objects (bridges, dams, etc.)

Conclusions

Determine the quantitative rating of the site, using the scale below. Circle the number on the data sheet that best describes what you observe.

HABITAT ASSESSMENT   **SCORE** _____

4 (excellent)	bends present, 5–10 riffles in 10 meters, many snags
3 (good)	bends present, 1–2 riffles in 10 meters, some snags
2 (fair)	occasional bend, 1–2 riffles in 50 meters, few snags
1 (poor)	straight channel, riffles/pools absent, no snags

Discussion Questions

1. Does your site provide a variety of habitats for aquatic organisms?
2. What features did you observe that might be especially productive?
3. What characteristics might make this a less desirable habitat?
4. Can you think of any changes that could improve the habitat characteristics?

Phytoplankton and Macrophyte Activities

Overview

Phytoplankton are tiny plants floating in water. These include green algae, blue-green algae, flagellates, and diatoms. The phytoplankton community is the foundation of the food web in rivers, lakes, and wetlands. Phytoplankton are also important because they take up carbon dioxide and release oxygen to water that is used by other organisms.

More diverse populations of phytoplankton generally indicate a healthy river, lake, or wetland. Although they are not usually found in the water column of fast-moving rivers and streams, filamentous forms of algae and diatoms can be found on rocks. Slow-moving river reaches and impoundments, as well as lakes and wetlands, often have many phytoplankton in the water column as well as growing on macrophytes, logs, and rocks.

Figure 7.15. How many bends do you see in this photo? Is this a young or an old river system? Are there any sources of pollution?

Phytoplankton communities depend upon dissolved nutrients (especially phosphates and nitrates) and sunlight. These communities respond to changes in nutrient levels, water temperature, turbidity, and pH.

High nitrate levels caused by inadequate sewage treatment, runoff of fertilizers from farms, runoff from feedlots, and livestock in the water contribute to a tremendous growth in blue-green algae (like Microcystis, Anabaena, and Nodularia). In streams with high nitrate levels, a thick carpet of filamentous algae may completely cover all of the rocks.

Figure 7.16. What might be the source of the plankton bloom noted in this photo? What effect does this bloom have on the water quality of the river?

Algal blooms and accumulations of floating scums are problems worldwide. Blooms or floating mats of blue-green algae may kill cattle, dogs, and other animals that drink from these waters; and humans can contract gastroenteritis, skin rashes, and other problems. Strong odors may also be produced by floating scums.

National Waterwatch, Australia, has provided leadership in the development of many innovative catchment, verge, riparian, and in-stream assessment approaches. Much of this section comes from their work.

Activity

(Optional)

Phytoplankton

Objective:

- Determine concentrations of blue-green algae.
- Identify blue-green algae and other kinds of algae.

Materials:

100 mL and 10 mL graduated cylinders, Lugols solution (see recipe in Appendix B), Sedgwick-Rafter Counting Cell, 1 mL pipette, and 400X microscope; no data sheets in Appendix B.

Time:

Approximately 1–3 hours.

Background Information

This activity is designed to determine concentrations of blue-green algae in a water sample. Blue-green algae is also present in healthy waters, but if an imbalance occurs in nutrient levels, blue-green algae easily dominate all other algae types. When the concentration of blue-green algae reach certain levels, then the water may pose a danger to livestock and humans. Determining the concentrations of blue-green algae provide some measure of relative risk to livestock and to humans who might have contact with the water.

Procedures

1. *Collecting a sample:*

   a. Collect a 1 liter sample from 10–20 cm below the surface (only in slow-moving rivers, impoundments, lakes, and wetlands).

   b. In streams, collect strands of filamentous algae from rocks or logs.

Figure 7.17. A 400X microscope may be needed to identify some of the common phytoplankton found in river systems.

2. *Preparing sample:*

   a. Examine the algal sample soon after it is taken.

   b. Thoroughly mix the sample and pour 100 mL of sample into a graduated cylinder.

   c. Add 1 mL of Lugols Solution with a pipette, for every 100 mL of sample. (Lugols solution is a preservative and a stain. Samples preserved in Lugols solution can be kept in the dark for up to a year.)

   d. Allow the sample to stand overnight to allow the phytoplankton cells to sink to the bottom of the graduated cylinder.

   e. The next day, gently pour off the top 90 mL of sample. A concentrated phytoplankton sample remains.

   f. Place 1 mL of the remaining sample into a Sedgwick-Rafter counting cell.

3. *Counting the phytoplankton:*

   The purpose of counting phytoplankton is determine concentrations in numbers/mL of sample. This concentration should represent the river, lake, or wetland from which it was taken.

The Sedgwick-Rafter counting cell is 50 mm long by 20 mm wide. The depth of the chamber is 1 mm. There are 50 squares or quadrants along the length of the counting cell and 20 squares or quadrants along the width of the counting cell. This totals 1000 quadrants. A strip represents 50 quadrants. A field represents one square.

Counting strips is somewhat less precise than counting fields. Which approach is chosen depends upon a determination of concentrations: High concentrations (more than 10 phytoplankton per field) are better calculated using the field counting approach; lower concentrations may best be calculated using the strip approach.

a. *Calculating numbers/mL phytoplankton by counting strips:*

$$\frac{Number}{mL} = \frac{C \times 1000}{L \times D \times W \times S}$$

where: C = number of organisms counted
L = length of each strip (cell length) in mm
D = depth of a strip (cell depth in mm
W = width of a strip (field image width) in mm
S = number of strips counted

b. *Calculating numbers/mL phytoplankton by counting fields:*

$$\frac{Number}{mL} = \frac{C \times 100 \text{ mm}^3}{A \times D \times F}$$

where: C = number of organisms counted
A = area of field (grid image area) in mm
D = depth of field cell depth in mm
F = number of fields counted

Conclusion

Determine the quantitative rating of phytoplankton, using the scale below. Circle the number on the data sheet that best describes what you observe.

PHYTOPLANKTON   **SCORE** _____

4 (excellent)	high diversity of phytoplankton (blue-green, green, diatoms, flagellates).
3 (good)	alert level I (500–2000 potentially toxic blue-green algal cells/mL).
2 (fair)	alert level II (2000–15000 potentially toxic blue-green algal cells/mL) concern for drinking water supplies.
1 (poor)	alert level III (greater than 15000 potentially toxic blue-green algal cells/mL) can cause death and illness in cattle and humans.

Refer to: Soil Conservation Service, United States Department of Agriculture. 1989. Water Quality Indicators Guide. Washington, D.C. or another local guide to the identification of local algae.

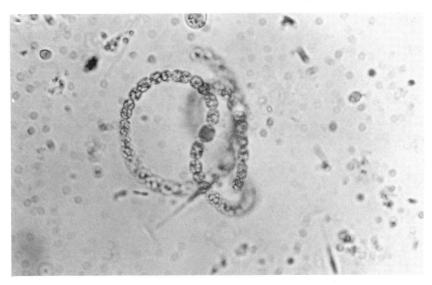

Figure 7.18. Some of the Anabaena algae are harmful in water supplies because they affect taste and odor and clog filters.

Discussion Questions

1. Do you feel that phytoplankton analysis can provide you with useful information in a river assessment program?
2. Does phytoplankton analysis provide any unique information?
3. Do you suspect that algae are seasonal or influenced by latitude?

Activity

(Recommended)

Macrophyte Cover

Objective:

- Assess the amount of macrophyte cover.

Materials:

Grid paper, measuring tape, no data sheets in Appendix B.

Time:

Approximately 1 hour.

Background Information

Aquatic macrophytes are large, often rooted plants, that live in the light zone of river pools, backwaters, and impounded areas. Few macrophytes inhabit fast-moving river reaches. Macrophytes serve important ecological functions within a river. They provide habitat for animals such as frogs, and fish serve as an anchor for the deposit of insect eggs, contribute organic matter to the food web, and help hold together the bottom and submerged banks. Altered flow, removal of riparian vegetation, increased turbidity, and greater nutrient levels affect the amount of macrophyte growth and the kinds of macrophytes found.

Procedure

1. Select a river reach that includes pools and riffles and that can be viewed across its width.
    a. Measure the width and length of river reach to be assessed. Calculate the scale on the grid paper.
    b. Observe the macrophytes growing in the reach.
2. Make a drawing of your stream to scale on grid paper.

Figure 7.19. Watercress plants normally indicate good water because they are found in cool, clear waters that flow all year.

3. In blocks of 100 squares mark on each grid an "M" (for macrophyte) for every quadrant that seems to contain a macrophytic plant.

4. Determine the percentage of "M" squares to total area.

5. Compare the percentage of macrophyte cover you find with the following quantitative rating scale.

Conclusions

Determine the quantitative rating of the surface and underwater plant cover using the scale below. Circle the number on the data sheet that best describes what you observe.

MACROPHYTE COVER   SCORE _____

4 (excellent)	patches of surface and underwater plant cover (<10 percent), abundant overhanging vegetation (excellent).
3 (good)	some surface and underwater plant cover (10–30 percent), some overhanging vegetation (good).
2 (fair)	abundant surface and underwater plant cover (10–50 percent), little overhanging vegetation (fair).
1 (poor)	choked surface and underwater plant cover (50–100 percent), no overhanging vegetation (poor).

Discussion Questions

1. Discuss your findings.
2. What other characteristics of your reach might contribute to the presence or absence of aquatic macrophytes?
3. What does the abundance of macrophytes you observed suggest about the water quality of the reach?

Benthic Macroinvertebrate Activities

Overview

Benthic macroinvertebrates are aquatic organisms that live on or in the substrate, or bottom, of a river or stream, and are large enough to be seen by the naked eye. Many characteristics of benthic macroinvertebrates make them useful indicators of water quality. They spend much of their life in the water, are easily sampled and identified, can be found worldwide, and are often sensitive to changes in water quality. Assessing the diversity of benthic organisms can provide a greater understanding of a river's condition.

This section provides instructions for benthic macroinvertebrate water quality activities. For greater detail on benthic organisms see Chapter 6 and Appendix A.

The methods used in the following activities generally follow a qualitative or semi-quantitative approach. They have been used successfully by non-biologists to monitor water quality and identify trouble spots. The activities include:

Activity 7.15   Sequential Comparison Index (SCI)—Recommended
Activity 7.16   Pollution Tolerance Index (PTI)—Recommended
Activity 7.17   EPT Richness—Optional

Note: Artificial substrates can be useful in situations where it is not possible to sample the natural substrate, such as in large rivers, channelized areas, or deep water. They can also provide standardized sampling. For details on how to use and make artificial substrates and other sampling devices for benthic organism collection and analysis, please see Chapter 6 and Appendix A.

Activity

7.15

(Recommended)

Sequential Comparison Index (SCI)

Objectives:

- Become more familiar with the range of taxa found in the river.
- Measure water quality by determining diversity or numbers of different kinds of benthic macroinvertebrates.

Materials and Equipment:

D-frame net or kick screen, waders, gloves, forceps, white tray, and wax pencil, paper, no data sheet in Appendix B.

Time:

Approximately 2 hours.

Background Information

The Sequential Comparison Index (SCI) is a measure of the distribution of individuals among groups of organisms.

$$SCI = \frac{\text{\# of runs}}{\text{total \# of organisms picked}}$$

This index relates to the diversity and relative abundance of organisms. This measure is easily used by people unfamiliar with benthic identification. The SCI is based on the theory of runs. A new run begins each time an organism picked from a sample looks different than the one picked just before it.

Procedures

1. Select sampling sites: The sites should be representative of the stream reach. Sampling should be done in a riffle area with a rubble or gravel bottom. Avoid areas below bridges, flow obstructions and artificial areas unless you are specifically testing for differences between these and other areas. If a good riffle area is not available, an artificial substrate may be placed in an area with visible flow.

2. Sample using a D-Frame net (Figure 7.20) or kick screen (Figure 7.21). To use a D-frame net, hold the opening of the net into the current and shuffle your feet upstream from the net. Benthic macroinvertebrates should be dislodged by your feet moving on the bottom and carried by the current into the net. A kick screen requires 2–3 people, one person holding each pole and a third person kicking the substrate upstream. Three samples, or at least 300 organisms, should be collected at each station. Spend the same length of time at each station.

3. Place the samples in 70 percent alcohol preservative for later sorting. Be sure to pick clinging organisms off the net (or at least a representative selection of them if you can't get them all). Keep the samples belonging to each station separate.

4. Following the instructions below, pick organisms from the sample to calculate the Sequential Comparison Index.

   a. Make a grid of 5–7 cm squares on the bottom of a white tray. (The grid may be laid-out with a permanent marker or wax pencil.) Number the squares in order.

   b. Rinse the sample of preservative, place it in the tray, and cover it with 0.5 cm–1 cm of water. Spread the sample evenly over the tray.

   c. Randomly select a starting grid from which to start picking the sample. Begin picking out organisms in a random sequence. Pick all specimens from one square before moving to the next. Continue picking until all, or 50 specimens are picked.

   d. Place organisms in a dish to compare each organism with the previously picked organism and record them on a work sheet using the symbols x and o (see example below). Record an "x" for the first organism picked. If the second organism picked is similar, record another "x." In the example below, the third organism picked is dissimilar to the previous organism, and so that is recorded as an "o," indicating a new run.

Example:

x x	o	x	o o o	x x	o
1	2	3	4	5	6

Total # runs = 6   Total individuals = 10

Resulting SCI = $\dfrac{\text{\# runs}}{\text{\# individuals}}$ = $\dfrac{6}{10}$ = 0.6

Figure 7.20. A D-Frame net may be useful in sampling the various "niches" in a river system. This might be a site for blackflies and net-building caddisflies.

e. After comparing specimens, place each in a petri dish containing similar organisms. This provides a rough sorting of the organisms into major groups to aid in identification.

f. To calculate the SCI, count the number of runs and divide by the total number of organisms.

g. Calculate an SCI for each sample. Average the samples to calculate a mean SCI for the site.

Conclusions

Determine the quantitative rating of the SCI, using the scale below. Circle the number on the data sheet that describes what you observed.

SCI VALUE   SCORE _____

4 (excellent)	0.9–1.0
3 (good)	0.6–0.89
2 (fair)	0.3–0.59
1 (poor)	0.0–0.29

Discussion Questions

1. It is thought that the riffle areas of a stream have the highest diversity. River bottoms of mud and silt most often support the lowest diversity. Which habitat has the highest SCI for your site?

2. To measure water quality differences along a river through the use of SCI requires comparing index values from similar substrates. What might cause differences in the SCI between two points?

Activity

 7.16

(Recommended)

Pollution Tolerance Index (PTI)

Objectives:

- Become more familiar with identification of organisms to order or family.
- Measure water quality through identification of organisms based upon pollution tolerance.

Materials and Equipment:

D-frame net or kick screen, waders, gloves, forceps, white tray, and wax pencil, Activity 7.16 data sheets in Appendix B.

Time:

 Approximately 2 hours.

Background Information

The Pollution Tolerance Index comes from "Save Our Streams" (The Izaak Walton League of America) and the "Citizen Stream Quality Monitoring Program" of the Ohio Department of Natural Resources. It is based on the concept of indicator organisms and tolerance levels. Indicator organisms are those organisms that are sensitive to water quality changes, and respond in predictable ways to changes in their environment. By their presence or absence they indicate something about water quality. The procedures are designed so that they can be done quickly and easily. This provides a rapid means of sampling riffle and other shallow areas in order to detect moderate to severe stream quality degradation.

Figure 7.21. This student is disturbing the stones in front of the "home-made" kick screen so the current will carry the aquatic animals into the screen.

The collected organisms are identified by comparing them with illustrations (Figures 6.10–6.25), or by using a key. Organisms are then classified into four groups based upon their pollution tolerance. Each of the four groups is given an index value, with the least tolerant group having the highest value. The Pollution Tolerance Index is determined by multiplying the number of kinds of organisms in each group by its index value; these numbers are then added together to form the Index. The general abundance of each kind of organism is also noted, although it is not figured into the Index.

Procedures

1. Choose a 3 meter by 3 meter area representative of the riffle or shallow area being sampled. Use the kick seine method (Figure 7.21) to sample this area. (If a seine is not available, several samples could be taken with a D-Frame net or kick screens.)

2. Three samples should be taken at each site to be sure a representative sample is collected. (Samples may also be taken from some of the other habitats at the site, such as on rocks in slow moving water, or near banks, since different organisms may be found there.)

3. Place samples in containers with a 70 percent alcohol solution for later identification. Be sure to pick clinging organisms off the net. (If you do not want to preserve the organisms, and if time allows, you may be able to identify and release them at the site.)

4. Record the presence of each type of organism collected and classify it by its tolerance (see table below). Estimate the number of each organism type (1–9, 10–49, 50–99, 100 or more) collected and record the appropriate scale on the evaluation sheet.

5. Calculate the Pollution Tolerance Index: multiply the number of types of organisms in each tolerance level by the index value for that level (4, 3, 2, or 1), and add the resulting four numbers. The following example demonstrates how to calculate an index for a hypothetical sample.

Group 1	Group 2	Group 3	Group 4
Caddisflies	Dragonflies	Blackflies	Tubifex
Stoneflies	Crayfish	Midges	Left-handed snails
Mayflies	Clams	Leeches	Blood Midge
3 x 4 = 12	3 x 3 = 9	3 x 2 = 6	3 x 1 = 3

Pollution Tolerance Index = (12+9+6+3) or 30

Cumulative Index Value	Stream Quality Assessment
23 and above	Excellent
17–22	Good
11–16	Fair
10 or less	Poor

Conclusions

Determine the quantitative rating of the PTI, using the scale below. Circle the number on the data sheet that best describes what you observed.

POLLUTION TOLERANCE INDEX (PTI)   SCORE _____

4 (excellent)	23 and above
3 (good)	17–22
2 (fair)	11–16
1 (poor)	10 or less

Discussion Questions

1. What information does a Pollution Tolerance Index provide that might be considered unique?

2. How reliable is a Pollution Tolerance Index?

Activity

(Recommended)

EPT Richness

Objectives:

- Consider Ephemeroptera, Plecoptera, and Tricoptera as indicators of good water quality.
- Note distinguishing physical characteristics among aquatic organisms within the same order.

Materials:

D-frame net or kick screen, waders, gloves, white pan, forceps, and wax pencil, no data sheets in Appendix B.

Time:

Approximately 1 hour to sample and identify to order.

Figure 7.22. Students analyzing their water with benthic macroinvertebrate data.

Background Information

EPT Richness is the number of taxa (based on gross physical differences) from each of the orders Ephemeroptera, Tricoptera, and Plecoptera. This measure is known to include many species that are sensitive to water quality changes. Generally, the more EPT taxa, the better the water quality.

Procedures

1. Choose a 3 meter by 3 meter area representative of the riffle or shallow area being sampled. Use the kick seine method to sample this area. (If a seine is not available several samples could be taken with a D-Frame net or kick screens.)

2. Three samples should be taken at each site to be sure a representative sample is collected. Samples may also be taken from some of the other habitats at the site—rocks in slow moving water, near banks—as different organisms may be found there.

3. Pick through the samples and count the number of taxa representing the three orders: Ephemeroptera, Plecoptera, and Tricoptera. The larger the number of taxa representing these three orders, the better the water quality.

Figure 7.23. Tweezers are useful to the professional in picking benthics off a rock, but for youth and the inexperienced investigators a spoon or scraper may be easier and less damaging to the organisms.

4. To derive an EPT/midge ratio simply divide the total number of EPT individuals by the total number of chironomid individuals. A healthy sample should show a ratio of at least 0.75.

Conclusions

The number of taxa, or EPT score, partially depends upon whether one sorts to order, family, genus or species. A count of 12 to 15 families indicates good water quality; a count of less than 7 or 8 is cause for concern.

It is important to use paired sites if possible when comparing sampling locations along the same river, or between rivers.

EPT RICHNESS   SCORE _____

4 (excellent)	more than 15 families
3 (good)	12–15 families
2 (fair)	8–12 families
1 (poor)	less than 8 families

Discussion Questions

1. What conclusions can you draw about your river through the use of the EPT rating?
2. Could the EPT rating be used universally?
3. What are the limitations of the EPT rating?

Physical and Chemical Water Quality Monitoring Activities

Introduction

People throughout the world have begun to look at water quality in their catchments as a tool to understand the relationship between land-use practices and the health of the environment. Outside of North America and Europe, however, the financial resources needed to purchase commercial water quality monitoring kits are often prohibitive. The nine physical-chemical water quality tests recommended by the National Sanitation Foundation (dissolved oxygen, fecal coliform, pH, BOD, temperature, total phosphates, nitrates, turbidity, and total solids) cost approximately US $500. This cost is beyond the financial means of many communities. A number of these tests, however, can be performed with less costly local materials.

Financial restrictions, the proposed use of the data, the skill level of the participants, and level of accuracy required, will help determine the funds needed to perform the nine water quality tests. These nine tests reflect the physical and chemical characteristics of the stream. Evaluating benthic invertebrates communities is also an effective indicator of water quality and can be performed at little or no cost. Benthic invertebrate populations often reflect water quality conditions. For example, the Ohio Department of Natural Resources (USA) performed a river study using biological and chemical assessments. In 6 percent of these rivers, the chemical tests indicated impairment, but the biological assessment indicated no impairment. In 36 percent of streams, the chemical evaluation indicated no impairment, while the biological assessment indicated that the stream was impaired. In 58 percent of tests, the chemical assessment and biological assessment were in agreement, suggesting that biological monitoring is likely to give an inexpensive, yet representative view of the overall health of a river (U.S. Environmental Protection Agency, 1990).

Although the accuracy of some low cost methods may be identical to the more costly kits, teachers may find that the low cost kits present some

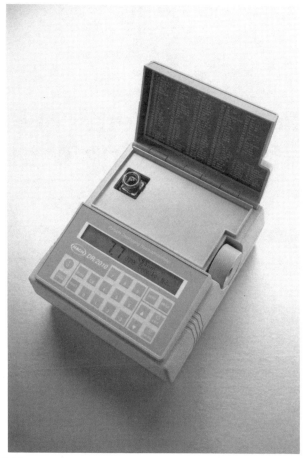

Figure 8.1. A DR/2010 spectrophotometer is a valuable piece of equipment to measure low-levels of heavy metals and other elements, but may be too costly for a school to purchase. It may be available at a local university. Photo courtesy of Hach Company.

problems. For example, Patrick Christie, a GREEN researcher assisting in a community water quality monitoring project in rural Nicaragua, provided both low-cost and commercial kits to students. He found students preferred to use the commercial equipment because they perceived the kits to be the more modern method for water monitoring. Local leaders and government officials had a similar perception, tending to have greater confidence in the more professional-looking kits. Christie also found the chemicals required for the inexpensive kits were difficult to obtain in Nicaragua.

Community involvement is an important variable for the success of a water quality monitoring program. By providing materials and expertise, communities can help water quality monitoring projects reduce expenses and build working relationships between students and professionals.

Places to look for water quality monitoring resources include hospitals (often have non-contaminated syringes and rubber gloves), technical schools (skills and materials to construct turbidity testing tubes and fecal coliform incubators), and universities or waste water treatment facilities (technical support or other assistance). Students gain from these collaborations by having the opportunity to work with environmental professionals such as chemists, water treatment engineers, and university professors. The following section addresses the costs and benefits of a number of water quality monitoring techniques. A list of equipment, cost and accuracy is included in Appendix A.

Developing a
Water Quality Monitoring System

Costs for water monitoring equipment are relative to the intended user. A fecal coliform vacuum assembly, which is expensive for a US high school, is low-cost to a wastewater treatment lab. Inexpensive items for American students, such as pH testing papers, may be very expensive for Third World students. Tests that a water quality investigator selects can be tailored to meet both the needs and budgets of almost any monitoring program. In this chapter, the following three options are identified for user consideration:

➤ **Option One:**
Involves purchasing commercial kits for each test. These kits can be refilled with chemicals purchased from supply companies. Commercial kits are available in a wide range of costs and accuracy, and are listed in Appendix A. This option is generally the most expensive but requires the least amount of time and effort to initiate a water quality monitoring program.

➤ **Option Two:**
Involves using a combination of commercial and local equipment for each test. For example, the refill chemicals for some kits can be purchased and used with glassware available in most chemistry labs.

➤ **Option Three:**
Involves very low-cost commercial kits, and/or materials obtained locally to produce school-made kits, nets and other equipment. This option requires the most preparation time but is the least expensive method to initiate a water quality monitoring program.

Users have some flexibility to design and use each physical-chemical test kit. For tests included under Option Two, glassware and purchased chemicals can be obtained from a variety of chemical companies or obtained locally. For information on commercial kits, refer to Appendix A.

Physical-Chemical Data: Nine Water Quality Tests

Overview

The National Sanitation Foundation selected 142 people, representing a wide range of positions, to develop a set of inclusive tests to determine water quality. Through a series of questionnaires, each person was asked to consider 35 tests. This number was reduced finally to a set of nine tests considered to be the most important to determine the water quality in river systems. These tests are listed below, along with an indication of under which testing option they may be performed:

Activity 8.1.	Dissolved Oxygen (1,2,3)
Activity 8.2.	Fecal Coliform (1,2,3)
Activity 8.3.	pH (1,2,3)
Activity 8.4.	Biochemical Oxygen Demand (BOD 5-day) (1,2,3)
Activity 8.5.	Temperature (3)
Activity 8.6.	Total Phosphate (1,2,3)
Activity 8.7.	Nitrates (1,2,3)
Activity 8.8.	Turbidity (1,2,3)
Activity 8.9.	Total Solids (1)

Activity

Dissolved Oxygen

(Recommended)

Objectives:

- Understand the significance of dissolved oxygen to water quality.
- Measure for dissolved oxygen.

Materials:

Materials and equipment as noted for each option (Appendix A); Activity 8.1 data sheets (Appendix B).

Time:

Approximately 20 minutes

Background Information

Dissolved oxygen (DO) is essential for the maintenance of healthy water systems. The presence of oxygen in water is an indication of good water quality, and the absence of oxygen is a signal of severe pollution. Rivers range from high to very low levels of dissolved oxygen—so low, in some cases, they are practically devoid of aquatic life.

Aquatic animals need oxygen to survive. Fish and some aquatic insects have gills to extract oxygen from the water. Some aquatic organisms, like pike and trout, require medium-to-high levels of dissolved oxygen to live. Other animals, like carp and catfish, flourish in waters of low dissolved oxygen. Waters of consistently high dissolved oxygen are usually considered healthy and stable ecosystems capable of supporting different kinds of aquatic organisms.

Generally, rivers with a constant dissolved oxygen saturation value of 90 percent or above are considered healthy, unless the waters are supersaturated due to cultural eutrophication. Rivers below 90 percent saturation may have large amounts of oxygen-demanding materials (organic wastes).

Sudden or gradual depletion in dissolved oxygen can cause major shifts in the kinds of aquatic organisms from pollution intolerant species to

I Figure 8.2. Rivers need to be healthy to support recreational activities.

pollution tolerant species. With a drop in dissolved oxygen levels, benthic communities containing many different kinds of aquatic insects sensitive to low dissolved oxygen—mayfly nymphs, stonefly nymphs, caddisfly larvae, and beetle larvae—might be reduced to a few different kinds, such as aquatic worms and fly larvae that are tolerant of these levels. (See Chapter 6 for more information on these and other organisms.) Nuisance algae and anaerobic organisms (that live without oxygen) may also become abundant in waters of low dissolved oxygen.

Methods for Measuring Dissolved Oxygen

➤ **Option One:**
Dissolved oxygen data can be obtained by using commercial kits distributed by a number of different chemical companies. The cost of dissolved oxygen kits are generally between US $30-$40. These kits include chemicals, glassware, and instructions.

➤ **Option Two:**
In addition to reagents, manufactured dissolved oxygen kits usually include a carrying case, a 60 mL sampling bottle, a measuring tube, a mixing bottle, and sometimes an eye dropper, all of which can be found in many chemistry labs. You can purchase the refill chemicals from the equipment manufacturers and run the test with your own glassware, a reduced-cost method which is just as accurate as a commercial kit. The

cost of the refill chemicals for the kits is between US $15 and $32 per 100 refills, a significant savings over commercial kits.

➤ **Option Three:**
If the cost of purchasing the dissolved oxygen chemicals remains prohibitive, the third option is to produce the reagents using chemicals that are available from a chemistry lab. These chemicals may be difficult for students to obtain and mix, but a chemistry teacher would have the background to combine chemicals to make the reagents. A scale that is accurate to 1/1000 of a gram is necessary. In addition, the chemicals must be mixed in quantities that are beyond the needs of a single water quality testing group. The reagents are usually mixed to volumes exceeding 100 mL, but only a fraction of a milliliter is needed to perform each test. To effectively utilize laboratory time and resources and prevent waste (some of the chemicals have only a two-week shelf life), networking among schools or groups on the same river will facilitate the use of these chemicals during a brief period of time.

Procedures for testing dissolved oxygen are in Appendix A.

Calculating Percent Saturation of Dissolved Oxygen

The percent saturation of water with dissolved oxygen at a given temperature is determined by pairing temperature of the water with the dissolved oxygen value, after first correcting your dissolved oxygen measurement for the effects of atmospheric pressure. This is done with the use of the correction table in Figure 8.4, and the percent saturation chart in Figure 8.5.

To calculate percent saturation, first correct your dissolved oxygen value (milligrams of oxygen per liter) for atmospheric pressure. Turn to Figure 8.4. Using either your atmospheric pressure (as read from a barometer) or your local altitude (if a barometer is not available), read across to the right hand column to find the correction factor. Multiply your dissolved oxygen measurement by this factor to obtain a corrected value.

Now turn to the chart in Figure 8.5. Draw a straight line between the water temperature at the test site and the corrected dissolved oxygen measurement, and read the saturation percentage at the intercept on the sloping scale.

Example

Let's say that your dissolved oxygen value was 10 mg/L, the measured water temperature was 15°, and the atmospheric pressure at the time of sampling was 608 mm Hg. From the table in Figure 8.4, the correction factor is 80 percent, which multiplied by 10 mg/L gives a corrected dissolved oxygen value of 8 mg/L. Drawing a straight line between this value and 15° gives a saturation of about 80 percent.

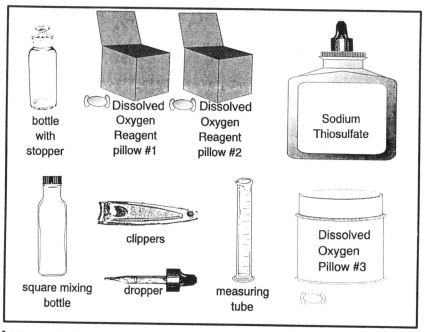

Figure 8.3. Reagents and glassware found in a commercial dissolved oxygen kit.

Atmospheric Pressure (mmHg)	Equivalent Altitude (ft.)	Correction Factor
775	540	1.02
760	0	1.00
745	542	.98
730	1094	.96
714	1688	.94
699	2274	.92
684	2864	.90
669	3466	.88
654	4082	.86
638	4756	.84
623	5403	.82
608	6065	.80
593	6744	.78
578	7440	.76
562	8204	.74
547	8939	.72
532	9694	.70
517	10,472	.68

Figure 8.4. Altitude correction table for dissolved oxygen measurements.

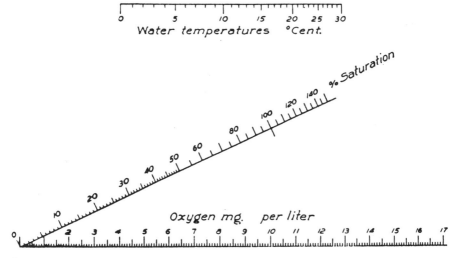

I Figure 8.5. Level of oxygen saturation chart.

How might you interpret these results? At the relatively cool temperature of 15°, one would expect a river to have a dissolved oxygen value higher than 80 percent. It would appear that something is using up oxygen in the water.

Generally, rivers that consistently have a dissolved oxygen value of 90 percent or higher are considered healthy, unless the waters are supersaturated due to cultural eutrophication. Rivers below 90 percent saturation may have large amounts of oxygen-demanding materials, i.e. organic wastes.

Conclusions

Determine the quantitative rating of the percent saturation of dissolved oxygen, using the scale below.

DISSOLVED OXYGEN (PERCENT SATURATION)   SCORE _____

4 (excellent)	91–110
3 (good)	71–90; > 110
2 (fair)	51–70
1 (poor)	< 50

Discussion Questions

1. What might cause a section of a river to become super saturated?

2. Why might this condition result in lower water quality?

3. Why is dissolved oxygen considered the most important of the nine water quality tests discussed in this chapter?

Activity

(Optional)

Fecal Coliform

Objectives:

- Understand the significance to water quality of fecal coliform.
- Measure for fecal coliform.

Materials:

Materials and equipment as noted for each option (Appendix A); Activity 8.2 data sheets (Appendix B).

Time:

Approximately 30 minutes.

Background Information

Fecal coliform bacteria are found in the feces of humans and other warm-blooded animals. These bacteria can enter rivers through direct discharge from mammals and birds; from agricultural and storm runoff carrying mammal and bird wastes; and, from sewage discharged into the water.

Fecal coliform bacteria naturally occur in the human digestive tract, and aid in the digestion of food. Fecal coliform bacteria by themselves are not pathogenic. Pathogenic organisms include bacteria, viruses, and parasites that cause diseases and illness. In infected individuals, fecal coliform bacteria occur along with pathogenic organisms.

If fecal coliform counts are high (over 200 colonies/100 mL water sample) in water, there is a greater chance that pathogenic organisms are present also. At this level, the probability of a person contracting a disease is great enough that a person should not be in contact with the water. The disease-causing organism could enter the human body through the nose, mouth, or ears, or through cuts in the skin. Diseases and illness, such as typhoid fever, hepatitis, gastroenteritis, dysentery, and ear infections can be contracted in waters with high fecal coliform counts.

Use	Standard (Colonies/100 ml)
Drinking Water	1 TC
Total Body Contact (Swimming)	200 FC
Partial Body Contact (Boating)	1000 FC
Treated Sewage Effluent	Not to exceed 200 FC

❙ **Figure 8.6.** General fecal coliform standards for different water uses.

Pathogenic microorganisms are relatively rare in water; this makes them difficult and time-consuming to monitor directly. Instead, because of the correlation between high fecal coliform counts and the probability of contracting a disease if directly exposed to the water, fecal coliform levels are monitored.

Methods for Measuring Fecal Coliform and Total Coliform Bacteria

➤ **Option One:**
Fecal coliform monitoring kits used by US and European schools are expensive. The pieces include: sterifil aseptic filtration system, hand vacuum pump assembly, plastic petri dishes with pads, S-Pak filters, plastic incubation bags, fecal coliform broth (or total coliform broth). The pieces must be ordered separately, and costs around $150. In addition, fecal coliform must be incubated at a constant temperature of 44.5° C for 24 hours. Commercial companies for obtaining the above equipment and supplies are found in Appendix A.

➤ **Option Two:**
All of the equipment and chemicals noted in "Option One" can be replaced. The major difference between fecal and total coliform monitoring is that fecal coliform are found only in the bodies of warm

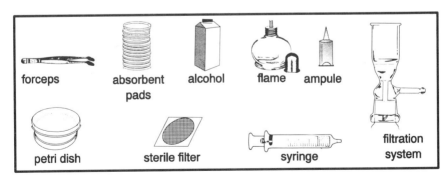

❙ **Figure 8.7.** Chemicals and equipment contained in a commercial fecal/total coliform kit.

Figure 8.8. A home-made incubator constructed by secondary students for incubating fecal coliform.

blooded organisms, whereas total coliform are found in the bodies of warm- and cold-blooded animals. Total coliform numbers generally run about 10 times higher than fecal coliform. Fecal coliform requires processing in an incubator, whereas total coliform can be incubated at room temperature. See Appendix A for ordering supplies and refills.

➤ **Option Three:**

Tupperware™ makes a low-cost fecal coliform filtration unit. It was originally intended to be a jelly mold. This piece of equipment is very inexpensive. The device may be ordered from Tupperware on all continents. The broth available from commercial firms is rather expensive, must be kept refrigerated, and has a shelf-life of only one year. OXOID membrane lauryl sulfate broth, a common laboratory ingredient, can be substituted for manufactured broth. A container of laurel sulfate costs $25 US for 100 grams. It can be pre-measured, then mixed with 3 mL of water to be used as a broth. See Appendix A for mixing chemicals, making a water bath, or obtaining low-cost equipment.

Procedures for testing fecal coliform are in Appendix A.

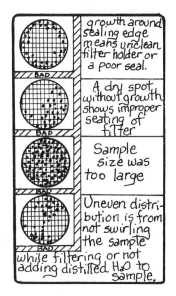

Figure 8.9. Obtaining a proper distribution of coliform in a culture.

Conclusions

Determine the quantitative rating of the fecal coliform, using the scale below.

FECAL COLIFORM (PER 100 ML)   SCORE _____
- 4 (excellent)   < 50 colonies
- 3 (good)   51–200 colonies
- 2 (fair)   200–1,000 colonies
- 1 (poor)   > 1,000 colonies

Discussion Questions

1. What is the origin of fecal coliform and total coliform?

2. What does fecal coliform indicate?

3. Some rivers of the world have fecal coliform levels of over one million/100mL of water—how would you run a sample on a river with extremely high fecal coliform levels?

Activity

(Recommended)

pH

Objectives:

- Understand the importance of pH to water quality.
- Measure of pH.

Materials:

Materials and equipment as noted for each option (Appendix A); Activity 8.3 data sheets (Appendix B).

Time:

Approximately 15 minutes.

Background Information

Water (H_2O) contains both H^+ (hydrogen) ions and OH^- (hydroxyl) ions. The pH test measures the H^+ ion concentration of liquids and substances. Each measured liquid or substance is given a pH value on a scale that ranges from 0 to 14. Pure, deionized water contains equal numbers of H^+ and OH^- ions and is considered neutral (pH 7), neither acidic or basic. If the sample being measured has more H^+ than OH^- ions, it is considered acidic and has a pH *less* than 7. If the sample contains more OH^- ions than H^+ ions, it is considered basic, with a pH *greater* than 7. See pH scale below.

It is important to remember that for every one unit change on the pH scale, there is approximately a ten-fold change in how acidic or basic the sample is. For example, lakes of pH 4 (acidic) are roughly 100 times more acidic than lakes of pH 6.

The pH of natural water is usually between 6.5 and 8.5, although wide variations can occur. Phenomena such as acid precipitation can alter the natural pH of a water body. Increased amounts of nitrogen oxides (NO_x) and sulfur dioxide (SO_2), primarily from automobile and coal-fired power plant emissions, are converted to nitric acid and sulfuric acid in the atmosphere. These acids combine with moisture in the atmosphere and fall to earth as acid rain or acid snow.

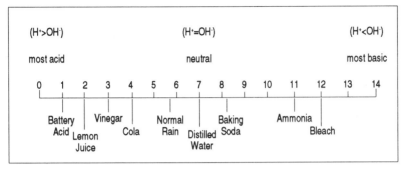

I Figure 8.10. Rain has a pH of 5.5 normally in the North America.

Changes in the pH value of water are important to many organisms. Most organisms have adapted to life in water of a specific pH and may die if the pH changes even slightly. At extremely high or low pH values (>9.6 or <4.5) the water becomes unsuitable for most organisms. Serious problems occur in lakes with a pH below 5, and in streams that receive a massive acid dose as acidic snow melts in the spring.

Methods for Measuring pH

➤ **Option One:**
The method many US students use for pH testing involves a color comparison between a colored disc and a control sample of water. Since the color comparison disc is an important feature of the kit, and the reagents sold by each company are specific for their colored discs, substituting homemade reagents is not applicable as with the dissolved oxygen test. These manufactured kits cost between $38 and $58 US and come with instructions.

Electronic pocket-sized electronic pH meters are available from an assortment of commercial companies. These battery-operated devices will operate for several hours on three 1.4 volt batteries, making the cost per test extremely inexpensive. The least expensive pH meter costs $30 US and is accurate to ±0.5 pH units. Prices of pH meters rise to $60 and higher.

➤ **Option Two:**
All of the chemicals, batteries and replacement parts are replaceable at a moderate-cost from many of the commercial distributors noted in Appendix A.

➤ **Option Three:**
Test strips are the least inexpensive way to test for pH. Using paper test strips requires no extra training or equipment. They are extremely inexpensive (US $4 for 200-300 tests) and can be less expensive by

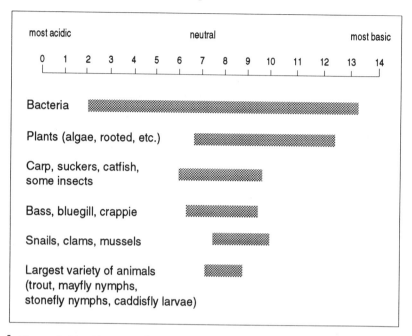

most acidic   neutral   most basic

0 1 2 3 4 5 6 7 8 9 10 11 12 13 14

Bacteria

Plants (algae, rooted, etc.)

Carp, suckers, catfish,
some insects

Bass, bluegill, crappie

Snails, clams, mussels

Largest variety of animals
(trout, mayfly nymphs,
stonefly nymphs, caddisfly larvae)

I Figure 8.11. pH ranges that support aquatic life.

cutting the strips in half lengthwise. Accuracy, however varies considerably on test strips. Some have errors as high as ±10 percent. Appendix A contains suppliers of pH test papers.

Procedures for testing pH are in Appendix A.

Conclusions

Determine the quantitative rating of the pH, using the scale below.

PH (UNITS)   SCORE _____

    4 (excellent)   6.5–7.5
    3 (good)   6.0–6.4; 7.6–8.0
    2 (fair)   5.5–5.9; 8.1–8.5
    1 (poor)   < 5.5; > 8.6

Discussion Questions

1. What might explain a river in an industrialized region having a pH level of 8.2?

2. Are aquatic insects equally affected by pH at all stages in their life cycle?

3. What impact does pH have on heavy metals in a river system?

4. What does pH stand for?

I Figure 8.12. Determining pH levels in water samples.

Activity

(Recommended)

Biochemical Oxygen Demand (BOD 5-Day)

Objectives:

- Understand the significance of Biochemical Oxygen Demand (BOD) to water quality.
- Measure BOD.

Materials:

Materials and equipment as noted for each option (Appendix A); Activity 8.4 data sheets (Appendix B).

Time:

Approximately 20 minutes for procedure, following a five-day incubation sample.

Background Information:

When organic matter decomposes, organic matter is broken down and oxidized (combined with oxygen) by microorganisms. Biochemical oxygen demand is a measure of the quantity of oxygen used by microorganisms in the aerobic oxidation of organic matter.

When aquatic plants die, they are fed upon by aerobic bacteria. In this process, organic matter is broken down and oxidized (combined with oxygen). Protozoa (like Paramecium) prey upon the growing population of bacteria and also require oxygen.

Decaying aquatic plants and their decomposers (aerobic bacteria), demand oxygen during plant decomposition. Nutrient input into the river—from nitrates and phosphates—will stimulate plant growth. Eventually, more plant growth leads to more plant decay. Therefore, nutrients can be a major force in high biochemical oxygen demand of rivers. Impounded river reaches also collect organic wastes from upriver that settle in quieter water; the bacteria that feed on this organic waste consume oxygen. Percent saturation (dissolved oxygen) values in waters with high plant growth and decay often fall below 90 percent.

In impounded and polluted rivers, much of the available dissolved oxygen is consumed by aerobic bacteria, robbing other aquatic organisms of the dissolved oxygen they need to live. Organisms that are more tolerant of lower dissolved oxygen, such as carp, midge larvae, and sewage worms, may become more numerous. Other organisms that are intolerant of low dissolved oxygen levels, such as caddisfly larvae, mayfly nymphs, and stonefly nymphs, will not survive in this water. In waters of high biochemical oxygen demand, a low diversity of aquatic organisms will replace the ecologically stable and complex relationships present in waters containing a high diversity of organisms.

Methods for Measuring Biochemical Oxygen Demand (BOD)

➤ **Option One:**
Testing for Biochemical Oxygen Demand requires the measurement of dissolved oxygen at two different time periods. Commercial kits are distributed by a number of chemical companies noted in Appendix A. The cost of dissolved oxygen commercial kits are between US $30–$40. Instructions are included in all kits.

➤ **Option Two:**
In addition to reagents, commercial dissolved oxygen kits usually include a carrying case, a 60 mL sampling bottle, a measuring tube, a mixing bottle, and sometimes an eye dropper—all of which can be

found in many chemistry laboratories. You can purchase the refill chemicals from the equipment manufacturers and run the test with your own glassware, a reduced-cost method which is just as accurate as a commercial kit. The cost of the refill chemicals for the kits is between US $15 and $32 per 100 refills, a significant savings over the pre-manufactured kits.

➤ **Option Three:**
If the cost of purchasing the dissolved oxygen chemicals needed to determine the biochemical oxygen demand remains prohibitive, the third option is to manufacture the reagents using chemicals that are available from a chemistry laboratory. These chemicals may be difficult for students to obtain and produce, but a chemistry teacher would have the background to combine chemicals to make the reagents. A weighing scale that is specific to 1/1000 of a gram is necessary. In addition, the chemicals must be mixed in quantities that are beyond the needs of a single water quality testing group. The reagents are usually mixed to volumes exceeding 100 mL, but only a fraction of a milliliter is needed to perform each test. To effectively utilize laboratory time and resources and prevent waste (some of the chemicals have only a two-week shelf life), networking among testing groups will facilitate the use of these chemicals during a brief period of time.

Procedures for testing biochemical oxygen demand are Appendix A.

Conclusions

Determine the quantitative rating of the BOD, using the scale below.

BIOCHEMICAL OXYGEN DEMAND (MG/L)		SCORE _____
4 (excellent)	<2	
3 (good)	2–4	
2 (fair)	4.1–10	
1 (poor)	>10	

Discussion Questions

1. If the BOD you measured was high, what possible sources of organic matter could be causing it?

2. Is this a problem?

3. Would you expect the BOD level in the river to fluctuate during a 24 hour period?

Activity

(Recommended)

Temperature

Objectives:

- Understand the significance of temperature to water quality.
- Measure the temperature of your river reach.
- Calculate the temperature change along the river.

Materials:

Materials and equipment as noted for option three (Appendix A); Activity 8.5 data sheets (Appendix B).

Time:

Approximately 30 minutes.

Background Information

The water temperature of a river is very important for water quality. Many of the physical, biological, and chemical characteristics of a river are directly affected by temperature. For example, temperature influences:

➤ the amount of oxygen that can be dissolved in the water (cool water can hold more oxygen than warm water, because gases are more easily dissolved in cool water);

➤ the rate of photosynthesis by algae and larger aquatic plants;

➤ the metabolic rates of aquatic organisms; and

➤ the sensitivity of organisms to toxic wastes, parasites, and diseases.

One of the most serious ways that humans change the temperature of rivers is through thermal pollution. Thermal pollution is an increase in water temperature caused by adding relatively warm water to a body of water. Industries, such as nuclear power plants, may cause thermal pollution by discharging water used to cool machinery for industrial processes.

Thermal pollution may also come from stormwater running off warmed urban surfaces, such as streets, sidewalks, and parking lots.

People also affect water temperature by cutting down trees that help shade the river, exposing the water to direct sunlight.

Soil erosion can also contribute to warmer water temperatures. As discussed in Chapter 4, soil erosion can be caused by many types of activities, including the removal of streamside vegetation, overgrazing, poor farming practices, and construction. Soil erosion raises water temperatures because it increases the amount of suspended solids carried by the river, making the water cloudy (turbid). Cloudy water absorbs the sun's rays, causing water temperature to rise.

As water temperature rises, the rate of photosynthesis and plant growth also increases. More plants grow and die. As plants die, they are decomposed by bacteria that consume oxygen. Therefore, when the rate of photosynthesis is increased, the need for oxygen in the water (BOD) is also increased.

The metabolic rate of organisms also rises with increasing water temperatures, resulting in an even greater oxygen demand. The life cycles of aquatic insects tend to speed up in warm water. Animals that feed on these insects can be negatively affected, particularly birds that depend on insects emerging at key periods during their migratory flights.

Most aquatic organisms have adapted to survive within a range of water temperatures. Some organisms prefer cooler water, such as trout and stonefly nymphs, while others survive under warmer conditions, such as carp and dragonfly nymphs. As the temperature of a river increases, cool water species will be replaced by warm water organisms. Few organisms can tolerate extremes of heat or cold, or rapid changes in water temperatures.

Temperature also affects aquatic life's sensitivity to toxic wastes, parasites, and disease. For example, thermal pollution may cause fish to become more vulnerable to disease, either due to the stress of rising water temperatures or the resulting decrease in dissolved oxygen.

The following temperature test measures the change in water temperature between two points in the river reach—the test site and a site 1.2 kilometers upstream. By detecting significant temperature changes along a section of the river, you can begin to uncover the sources of thermal pollution.

► **Option Three:**

The temperature test is extremely easy to perform. Thermometers can be purchased at a reasonable price. Because standard equipment is available at such low cost, this test is designated Option Three.

Procedures for testing water temperature are in Appendix A

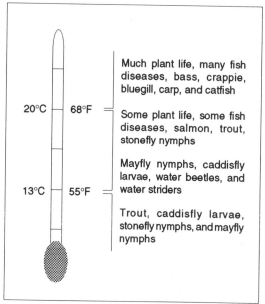

Figure 8.13. Temperature tolerance levels of selected aquatic organisms.

Conclusions

Determine the quantitative rating of the temperature difference between the two sites, using the scale below.

TEMPERATURE (Δ IN DEGREES CELSIUS)   SCORE _____

4 (excellent)	0–2
3 (good)	2.2–5
2 (fair)	5.1–9.9
1 (poor)	10 >

Discussion Questions

1. If the temperature difference was great, what possible sources of thermal pollution might contribute?

2. What factors might make your measurements inaccurate?

3. If measuring water temperature between two points, what might create a lowering of water temperature in a one mile stretch of river?

Activity

(Optional)

Total Phosphate

Objectives:

- Understand the significance of phosphate to water quality
- Measure of total phosphate in your river reach.

Materials:

Materials and equipment as noted for each option (Appendix A); Activity 8.6 data sheets (Appendix B).

Time:

Approximately 30 minutes.

Background Information

Phosphorous is usually present in natural waters as phosphate (PO_4-P). Total Phosphates includes organic phosphorus and inorganic phosphate. Organic phosphate is a part of living plants and animals, their by-products, and their remains. Inorganic phosphates include the ions ($H_2PO_4^-$, $HPO_4^=$, and $PO_4^\equiv$) bonded to soil particles, and phosphates present in laundry detergents.

Phosphorus is an essential element for life; it is a plant nutrient needed for growth, and a fundamental element in the metabolic reactions of plants and animals. In most waters, phosphorus functions as a "growth-limiting" factor because it is present in very low concentrations. This scarcity of phosphorus can be explained by its attraction to organic matter and soil particles. Any unattached or "free" phosphorus, in the form of inorganic phosphates, is rapidly taken up by algae and larger aquatic plants. Because algae only require small amounts of phosphorus to live, excess phosphorus causes extensive algal growth called "algal blooms." Algal blooms color the water a pea-soup green and are a classic symptom of *cultural eutrophication.*

Cultural eutrophication is an enrichment of the water, usually by phosphorus, from human activities. Natural eutrophication is rare. Forest

---

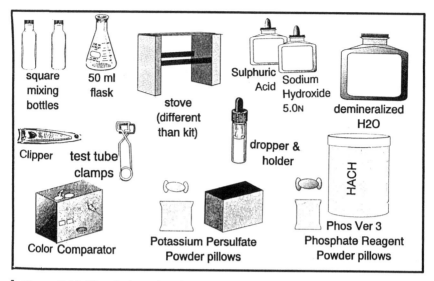

Figure 8.14. Chemicals and equipment in a commercial total phosphate (PO_4-P) kit.

fires and fallout from volcanic eruptions are natural occurrences that release phosphorus and may cause eutrophication. Lakes that receive no inputs of phosphorus from human activities age very slowly.

Water in advanced stages of cultural eutrophication can become anaerobic (without oxygen). Anaerobic conditions usually occur near the lake or impoundment bottom and produce gases like hydrogen sulfide which foul the shoreline with a "rotten egg" smell.

Methods for Measuring Total Phosphates

➤ **Option One:**
The total phosphate kits used by American and European schools costs approximately $100 US. This kit comes with a case, assorted glassware, a small stove, and all of the necessary chemicals. Instructions in Appendix A.

➤ **Option Two:**
All chemicals provided in a commercial total phosphate kit are replaceable.

➤ **Option Three:**
Produce your own total phosphate kit. The chemicals could be prepared in a school chemistry laboratory. The glassware would generally be available in a science laboratory. An available gas fired camp stove could replace the stove included in the kit. Instructions in Appendix A.

Procedures for testing total phosphates are in Appendix A.

Conclusions

Determine the quantitative rating of the total phosphates, using the scale below.

TOTAL PHOSPHATE (MG/L)   SCORE _____
4 (excellent)	0–1
3 (good)	1.1–4
2 (fair)	4.1–9.9
1 (poor)	> 10

Discussion Questions

1. Was your total phosphate level high or low and what might explain the results?
2. Does total phosphate effect dissolved oxygen readings?
3. What might explain a high phosphate level in a mountain stream?

Activity

(Optional)

Nitrates

Objectives:

- Understand the significance to water quality of nitrates.
- Measure for nitrates in your river reach.

Materials:

Materials and equipment as noted for each option (Appendix A); Activity 8.7 data sheets (Appendix B).

Time:

Approximately 20 minutes.

Background Information

Nitrogen is an essential plant nutrient required by all plants and animals for building protein. In aquatic ecosystems, nitrogen is present in many different forms.

Nitrogen is most abundant in its molecular form (N_2) which makes up 79 percent of the air we breathe. It is a much more abundant nutrient than phosphorus in nature. In its molecular form (N_2), nitrogen is useless for most aquatic plant growth.

Blue-green algae, the primary component of algal blooms, are able to use the molecular form of nitrogen (N_2) and biologically convert it to usable forms of nitrogen for aquatic plant growth, ammonia (NH_3) and nitrates (NO_3^-).

Aquatic animals obtain nitrogen by eating aquatic plants and converting plant proteins to specific animal proteins or by eating other aquatic organisms which feed upon plants.

Because nitrogen, as ammonia and nitrates, acts as a plant nutrient, it also causes eutrophication. Eutrophication causes more plant growth and decay, which in turn stimulates a biochemical oxygen demand. However, unlike phosphorus, nitrogen rarely limits plant growth, so plants are not as sensitive to increases in ammonia and nitrate levels.

Sewage is the main source of nitrates added by humans to rivers. Sewage enters waterways in inadequately treated wastewater from sewage treatment plants, in the effluent from illegal sanitary sewer connections, and from poorly functioning septic systems. Other important sources of nitrates in water are fertilizers, and the runoff from cattle feedlots, dairies, and barnyards. High nitrate levels have been discovered in groundwater beneath croplands due to excessive fertilizer use, especially in heavily irrigated areas with sandy soils.

Water containing high nitrate levels can cause a serious condition called *methemoglobinemia* (met-hemo-glo-bin-emia), if it is used for infant milk formula. This condition prevents the baby's blood from carrying oxygen.

Methods for Measuring Nitrates

➤ **Option One:**
Commercial nitrogen kits can be obtained from a number of worldwide distributors (see Appendix A). The cost of a commercial nitrate kit is around US $35-60.

Figure 8.15. Students drawing a nitrate sample by using a home-made extension tube.

➤ **Option Two:**
All of the chemicals provided in a commercial nitrate kit are replaceable. See Appendix A.

➤ **Option Three:**
Nitrate test strips, similar to those used for pH, are available from various distributors (see Appendix A). The strips are swirled in water, removed and compared with a color code, and read in milligrams per liter.

Procedures for testing nitrates are in Appendix A.

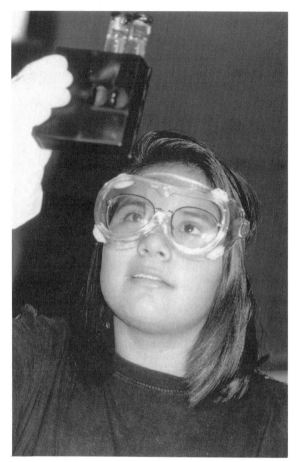

Figure 8.16. A Native American youth checking the waters on her reservation for nitrates using a commercial NO_3 kit.

Conclusions

Determine the quantitative rating of the nitrates, using the scale below.

NITRATES (MG/L)   SCORE _____

    4 (excellent)   0–1

    3 (good)   1.1–3

    2 (fair)   3.1–5

    1 (poor)   > 5

Discussion Questions

1. If you found a high level of nitrates in your river reach, what are possible sources?
2. Is there a correlation between nitrate readings and dissolved oxygen readings?

Activity

Turbidity

(Recommended)

Objectives:

- Understand the significance to water quality of turbidity.
- Measure of turbidity in your reach.

Materials:

Materials and equipment as noted for each option (Appendix A); Activity 8.8 data sheets (Appendix B).

Time:

  Approximately 15 minutes.

Background

Turbidity is a measure of the relative clarity of water: the greater the turbidity, the more "murky" the water. Turbidity increases as a result of suspended solids in the water that reduce the transmission of light. Suspended solids are varied, ranging from clay, silt, and plankton, to industrial wastes and sewage.

Turbidity may be the result of soil erosion, waste discharge, urban runoff, abundant bottom feeders of ship traffic that stir up bottom sediments, or the presence of excess nutrients that result in algal growth. Turbidity may affect the color of the water, from nearly white to red-brown, as well as green from algal blooms.

Figure 8.17. Students using a home-made turbidity tube calibrated with a commercial turbidimeter by Australians.

At higher levels of turbidity, water loses its ability to support a diversity of aquatic organisms. Water becomes warmer as suspended particles absorb heat from sunlight, causing oxygen levels to fall (remember, warm water holds less oxygen than cooler water). Less light penetrating the water decreases photosynthesis, causing further drops in oxygen levels. The combination of warmer water, less light, and oxygen depletion makes it impossible for some forms of aquatic life to survive. In addition, suspended solids may clog fish gills, reduce growth rates, and decrease resistance to disease, as well as prevent egg and larval development.

Methods for Measuring Turbidity

➤ **Option One:**
Due to the physical nature of this test, there are numerous ways that turbidity can be analyzed. In the U.S., some watershed programs have purchased a specialized instrument called a turbidimeter, which measures the amount of light scattered through a sample of collected water. A turbidimeter is very accurate, but costs over US $600. The turbidimeter comes with precise instructions and reads measurements in Nephelometer Turbidity Units (NTUs). See Appendix A for locating commercial distributors.

➤ **Option Two:**
LaMotte has a relatively inexpensive "Turbidity Kit" (US $29.50) that compares the turbidity of a measured amount of the sample with an identical amount of turbidity-free water containing a measured amount of standardized turbidity reagent. The measurements are in Jackson Turbidity Units (JTUs) which are interchangeable units with Nephelometer Turbidity Units NTUs. See Appendix A for instructions.

The Australian Waterwatch Program has designed a "Turbidity Tube" (77.5 cm/30") that sells for under US $30. The tube is calibrated to read in NTUs. See Appendix A for instructions.

➤ **Option Three:**
Turbidity can be measured by a *Secchi disk*, which is inexpensive and easily made by students (Figure 1.4, chapter 1). However, a Secchi disk can not be used in shallow water or in a strong current. See Appendix A for instructions for making a Secchi disk.

Problems may occur using a Secchi disk. If the river is too shallow and one can see the bottom, the Secchi disk cannot be lowered deep enough to measure turbidity. Strong currents can angle the Secchi disk downstream, making the reading inaccurate. Attempting to correct this with heavy weights can make the disk difficult to raise and lower. See Appendix A for making a Secchi disk.

Conclusions

Determine the quantitative rating of the turbidity, using the "turbidity tube" scale below. Circle the number on the data sheet that describes your data.

TURBIDITY			SCORE _____
Rating	NTUs	Secchi disk	
4 (excellent)	0–10	>3 feet	
		>91.5 cm	
3 (good)	10.1–40	1 foot to 3 feet	
		30.5 cm to 91.5 cm	
2 (fair)	40.1–150	2 inches to 1 foot	
		5 cm to 30.5 cm	
1 (poor)	> 150	< 2 inches	
		< 5 cm	

Discussion Questions:

1. What factors might contribute to the turbidity levels you found?

2. How might turbidity readings effect dissolved oxygen readings?

Activity

(Optional)

Total Solids

Objectives:

- Understand the significance of total solids to water quality.
- Measure of total solids in your river reach.

Materials:

Materials and equipment as noted for each option (Appendix A); Activity 8.9 data sheets (Appendix B).

Time:

Approximately 24 hours.

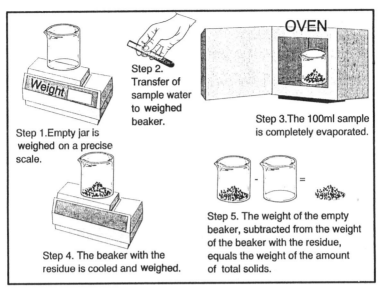

Figure 8.18. Procedure and equipment for testing the river for total solids.

Background Information

Total solids, also referred to as total residue, is a water quality measure of: (1) dissolved solids or that portion of solid matter found in a water sample that passes through a filter; and (2) suspended solids or that portion of solid matter that is trapped by a filter.

Dissolved inorganic materials include calcium, bicarbonate, nitrogen, phosphorus, iron, sulfur, and other ions found in a water body. Consistent concentrations of these minerals is essential for the maintenance of aquatic life. Also, many dissolved ions, such as nitrogen, phosphorus, and sulfur are building blocks of molecules necessary for life. Suspended solids include silt and clay particles from soil runoff, plankton, industrial wastes, and sewage.

High concentrations of dissolved ions cause water balance problems for individual organisms. Low concentrations may limit the growth of aquatic life, or restrict some organisms from surviving in the water.

High concentrations of dissolved solids in drinking water can lead to laxative effects in humans and impart an unpleasant mineral taste to the water. High concentrations of suspended solids also reduces water clarity, contributes to a decrease in photosynthesis, binds with toxic compounds and heavy metals, and leads to an increase in water temperature through greater absorption of sunlight by surface waters.

The total solids test is relatively easy to conduct if the proper equipment is available. The test requires a scale accurate to 1/10,000 of a gram. No low-cost options exist; however, the scale may be available through local organizations such as universities or laboratories.

Methods for Measuring Total Solids

➤ **Option One:**

Total solids is considered option one because it requires a scale accurate to 1/10,000 of a gram. A scale of this accuracy may be available at a local university, waste water treatment plant, or a local industry.

Procedures for testing total solids are in Appendix A.

Conclusions

Determine the quantitative rating of the total solids, using the scale below.

TOTAL SOLIDS (MG/L)		SCORE _____
4 (excellent)	< 100	
3 (good)	100–250	
2 (fair)	250–400	
1 (poor)	> 400	

Discussion Questions

1. If the total solids you measured were high, what are some possible sources of the material?

2. Do you think the dissolved solids or the suspended solids contributed more to total solids readings?

Action-Taking and Activities

Introduction

The previous chapters provide background information about catchments, detail approaches to observing the river and its catchment from a social, geographical, cultural and political viewpoint, and offer guidance in making biological, physical, and chemical measurements. But where do all of these observations and measurements lead? The purpose of this chapter is to help water quality programs move from observations and measurements to a definition of the problem and an exploration of solutions. First, some background and description of action-taking is explored. Next, a series of activities for moving through the action-taking process are explored. Finally, several case studies of water quality monitoring programs taking action are present. The three primary steps in an action-taking process are:

1. to move from observations and measurements to interpretation of the data and definition of the problem;
2. to explore solutions to the problem based upon accepted criteria and with successful outcomes in mind; and,
3. to take action and assess whether the desired outcomes been reached.

Environmental problems are by nature interdisciplinary, embracing science, social science, economics, and politics. Effective solutions to these problems rest upon an understanding of, and respect for, the culture in which they are found. Because environmental problems are interdisciplinary in scope, and often affect diverse populations with differing values and perspectives, effective solutions require looking at problems from various vantage points. The complexity of environmental problems, and appropriate responses to these problems, require a sense of competence that stems in part from a thorough knowledge of the problem area.

Figure 9.1. Students noting the natural movement of aquatic animals they had just placed in the container.

The Global Rivers Environmental Education Network's catchment monitoring model combines aspects of fieldwork focusing on rivers and catchments internationally, with a community problem solving approach that follows the three steps noted above. The model is based on research into how we organize ideas (mapping), how people solve problems, and how educators can use this information to make education more relevant and meaningful. Ultimate success is measured by improved river systems, thereby improving the lives of people dependent upon rivers.

Background

The process followed in taking effective action has been influenced by researchers in cognitive psychology, by environmental educators seeking effective means to educate and empower people about their environment, and by educational researchers interested in problem solving strategies within educational contexts.

In cognitive psychology, researchers have found that people form cognitive maps—ways of organizing information based upon experiences. These maps represent our understanding of the world. For example, children learn to categorize and recognize everyday objects like shapes. Circles come in the form of wheels on carts and as the sun or full moon. As children become older, higher levels of understanding, or higher level maps, mean that students in secondary schools can comprehend the abstraction of a circle of life as noted in natural systems and in some religions. Cognitive maps allow people to navigate life. They help people to assimilate information and give some ability to predict, and to anticipate some behaviors and events. When people come up against situations in which their cognitive maps do not fit, they seek to find a solution—to take action.

The educator John Dewey was an early advocate of reflective thinking, and believed that thoughts cannot be separated from action (Dewey, 1963). He thought that schools should involve students in the real world to make education more meaningful. Later, Kurt Lewin, a social scientist, worked with Stephen Corey at Columbia University to apply the principles of action research to educational settings. This approach linked theories of social change with action. Paulo Friere, a Brazilian educator, worked to empower illiterate people in Brazil. His philosophy is represented in the following statement (Friere, 1970):

> *You never really understand an issue or know how to help resolve it until you involve yourself in the issue. Then you begin to understand it, to identify the principal parties and actors involved, and begin to realize how to change it.*

Community problem solving consists of the following elements: recognizing a problem; setting problem objectives; working in groups; collecting, organizing, and analyzing information; defining the problem from different perspectives; identifying and selecting alternative actions; carrying out those actions; and evaluating outcomes and the process (Brody, 1982).

The employment of community problem-solving and action research in the area of environmental education has come most notably from the work of William Stapp, Arjen Wals, Giovanna Di Chiro, Ian Robottom, Bill Hammond, and Harold Hungerford. This is the educational philosophy upon which GREEN is founded.

Exploring and Defining the Problem

The observations of water quality made using this manual, and the measurements taken of benthic diversity, physical attributes like flow and discharge, and chemical indicators like nitrates, provide an overall assessment of the health of a river and its catchment. But these are observations and measurements of indicators and not of the problems themselves. For example, one may measure low benthic diversity. One interpretation of this measurement is that low dissolved oxygen levels may be the problem. Dissolved oxygen and its percent saturation levels in turn may indicate a

Figure 9.2. Heavy metals being extracted from old computers in large vats on the riparian area of a river in Taiwan—a potential source of pollution.

problem with enrichment of the water through over-fertilization, or from sewage discharges. The problem may be caused by a combination of factors: for example, over-fertilization of adjacent fields because of poor soil fertility caused by erosion of valuable topsoil over time and by farming in areas prone to erosion because of the pressures of overpopulation.

In defining a problem, it is important to focus attention on the interpretation of all the observations and measurements. Some questions that might be helpful in interpreting and defining a problem include:

➤ Do observations and measurements reveal patterns or trends?

➤ Do these observations and measurements seem to indicate a generalized problem?

➤ How is a problem defined? By impaired human usage of the water? By degradation of aquatic life?

➤ What are the root causes of the problem?

➤ Who is affected by the problem? Who is responsible for solving the problem?

➤ Is the problem temporal in nature? What are the factors creating or sustaining the problem?

The initial definition of a problem may be the most significant factor leading to effective action-taking. The following scenario illustrates the interconnectedness and complexity of environmental problems, and the information gathering, integrative aspects of problem definition.

A Scenario of a Catchment Monitoring Program

Students decided to conduct a project along a river reach in their community. Through background reading, they discovered the dynamic relationship between a river and its bank, and especially the role that riparian vegetation plays in erosion control, nutrient uptake, and protection of water quality. To gain a better understanding of the river and its relationship to the land, students chose to conduct a catchment assessment along a 1 km reach of river in their community (see Chapter 7).

They observed and recorded a mix of land uses: mostly agriculture with row crops and significant cattle pasturing; some abandoned businesses along the river; and some new housing being built away from the river. They found no mining along this reach, however, there were known to be bauxite mines in the upper reaches of the catchment. Water temperature seemed appropriate for the season. The bottom of the river was composed of mostly silt and soft substrates and river velocity was slow (less than 0.2 m/sec).

Figure 9.3. Students on a river walk to note land use practices along the riparian area and implications for water quality.

The river was channelized along the business area, and was altered by irrigation ditches entering the row crop areas and leading back to the river. The water served as the primary drinking water source for the community. People fished along the river. It was used for agriculture, as the industrial water supply, and for waste disposal. Cattle had access to the river.

The river was very muddy and was known to flood periodically. There were virtually no wetlands left along the observed reach—the bottom seemed fairly uniform. The banks were steep along much of the reach and denuded because of cattle. Native vegetation was severely disturbed, with extensive erosion on the banks leading to the river.

One Potential Problem

The students observed that many cattle had access to the river along this reach. Students also talked to village officials and found that the community takes in its drinking water along this reach. The river served as a drinking water source for both cattle and people—could there be a problem with cattle feces entering the water supply? This question prompted a study of total coliform levels near the water intake for the city. Total coliform tests were made because this is a less expensive test—an incubator isn't necessary—and because total coliform levels correlate with fecal coliform levels in a roughly 10 to 1 ratio (total to fecal). Very high total coliform levels were found, correlating to an approximate level of 2,000 fecal coliform colonies/100 mL. Students consulted a local non-governmental organization (NGO) and discovered that, according to the World Health Organization, acceptable levels were at 200 colonies fecal coliform for swimming and only 1 total coliform colony for drinking water.

The next question the students asked was: does the city adequately treat the water? Students found that the city used some chlorination during the warmer months, but no systematic testing was done after chlorination to determine if safe drinking water levels had been reached. In response, the students decided to measure total coliform levels in drinking water from several different areas of the community and compare results over time. They discovered that the chlorine did kill much of the coliforms, but that during the cooler rainy season drinking water showed unacceptable levels of coliforms. It was determined that cattle were likely the primary cause of high coliform levels because they had unrestricted access to the river.

A Second Potential Problem

The initial catchment assessment noted very muddy waters and severe erosion along the river bank. What was the cause of the erosion? Students decided to do a follow-up investigation recording activity along the riverbank and monitoring turbidity before and after rainfalls. They discovered that cattle coming down to the river had not only laid the banks bare, but had also caused severe erosion of the banks with their hooves.

In this scenario, the problems point to the same source—unrestricted access to cattle along the river. But the first problem led to another question: Should the community be chlorinating year-round? And should the community develop a systematic testing program for its drinking water? The cattle access problem might be phrased in this way: unrestricted access by cattle to the river has caused severe erosion of the river banks and resulting high turbidity levels, and has contributed significant loads of fecal coliform to the river. In attempting to develop solutions, students needed to ask more questions and gather more information. Why were cattle sent to the river for water to drink. Were there not alternative sources of water?

Exploring Solutions to Problems

In considering possible solutions to a defined problem or problems it is helpful to establish criteria in the beginning for acceptable solutions to work toward. Criteria might include: How much time would it take to achieve this solution? Who will be affected by the proposed solution? Do we have enough information (can we get enough information in the time we have) to achieve this? Do we care about this problem? Is it interesting? Is it significant?

Objectives should be tied to successful outcomes. A successful outcome might be: cattle will not have access to the river, leading to the rejuvenation of vegetation on river banks and a decline in coliform levels. But what steps must take place for this to occur? Who else might care about this problem or be responsible for it? There might have to be a survey of cattle farmers in the area to ask why cattle have unrestricted access. There may be economic constraints to limiting access—fences are costly to construct and alternative sources of water for the cattle may have to be developed.

The other problem uncovered during this problem-solving process was the inadequate treatment and testing of drinking water. Students might not have enough information to make the case that this is a problem. Students might have to survey households for incidence of diarrhea, gastroenteritis, hepatitis, and other diseases connected with contaminated water. If a case can be made to chlorinate more regularly, the cost may be prohibitive for the community. People could boil their water during the cooler rainy season when chlorination levels are lower, or they may already do so.

Figure 9.4. Bus trip taken to interview some of the land owners in the upper reaches of the Mary River catchment in Australia.

Taking Action

Taking action means taking steps toward meeting those objectives students have identified. In the above scenario, these steps could include an interview of landowners, an analysis of the costs of fence construction or alternative sources of water for cattle, presenting findings of their work to community officials, seeking letters of support from the officials responsible for drinking water and for restricting access of cattle, and then later assessment of the river bank for erosion and of total coliform levels as access is restricted.

Action also may occur at several levels: individual, school, and community. Using the above scenario, action-taking at the *individual* level might include planting trees and other vegetation along the river. At the *school* level, students who have researched the problem could enlist the help of other students and teachers in expanding the total coliform testing to include the school. At the *community* level, students could take some of the actions described above: interviewing landowners, seeking support from city officials, and conducting a cost-benefit type study of the means to control cattle and provide drinking water.

Effective action-taking may require a constant rethinking of strategies based upon constraints that may arise. If many potential constraints are identified early in the process, alternative strategies can be devised. The following set of activities is designed to help students along an action-taking path.

Action-Taking Activities

Rivers around the world face a variety of complex problems. Increased water demand by humans, deforestation, inadequate treatment of animal and human wastes, and toxic contamination all contribute to water quality problems.

It is important to learn how to confront these complex problems in creative ways. Solving problems and taking action can help students realize that their knowledge, talents, and efforts are worthwhile. Activities that encourage problem-solving and action-taking may be a very satisfying way to conclude the water quality monitoring program.

The activities in this section are designed to awaken student concern; develop tools to gather more information; build problem solving skills vital to the future; and instill a greater sense of confidence in their abilities. They are structured to help lead you through the process of problem-solving and action-taking. The activities include:

Activity 9.1   Identifying Specific Problems (Recommended)
Activity 9.2   Visualizing the Future (Recommended)
Activity 9.3   Selecting an Issue to Address (Recommended)
Activity 9.4   Contacting Organizations and Decision Makers (Recommended)
Activity 9.5   Developing an Action Plan (Recommended)
Activity 9.6   Taking Action (Recommended)
Activity 9.7   Follow-up (Recommended)

We recommend that you use all seven of these activities, in the order given, with student groups who plan to take action. However, if your class is not going to undertake a major action project, consider using Activity 9.2 Visualizing the Future, and Activity 9.3 Contacting Organizations and Decision Makers. These two activities are useful exercises by themselves, and provide an introduction to some problem-solving skills.

Activity

(Recommended)

Identifying Specific Problems

Objectives:

- Analyze and evaluate water quality information.
- Synthesize students' ideas and impressions.
- Identify root causes of water pollution.
- Build skills working in groups.

Materials:

Pencil, paper, blackboard or newsprint paper, completed data sheets from water quality monitoring activities.

Time:

Approximately 80 minutes.

Background Information

Once you have collected water quality and land use information from a variety of sources, you may be concerned about specific problems you have discovered. This activity gives you a chance to synthesize the information you have gathered and begin to ask, "What are the most serious threats to water quality? What are their root causes?"

Procedures

1. Divide into small groups for a brainstorming activity. Use all the information available (including your own test results, test results from other schools obtained through the computer conference, direct observations, the *Field Manual for Water Quality Monitoring*, etc.) to generate a list of problems that impact the river.

2. Once the groups have created their lists of problems, representatives from each group should share their lists with the rest of the class. Generate a master list of major issues, drawn from the groups' lists, on the blackboard or on a sheet of paper.

3. Next, review the list, and for each one discuss the problem in some detail.

   a. Distinguish between the more obvious, immediate cause of a problem, and the various levels of underlying root causes (which are usually not so obvious).

   For example, you may have identified a problem like a very high fecal coliform count at one spot along the river. From your field work, you can identify that the immediate, direct cause of this problem is a point source of organic pollution from a leaky sewer pipe. But what are some of the possible root causes of this problem? Students might conjecture that the local city government, responsible for repairing the pipe, has not done so due to budget constraints. This is one root cause of the pollution problem.

   b. If possible, identify other root causes of the problems, at a deeper level than the first one. For the example given above, the fact that the city cannot afford to fix the sewer pipe may be traced to the budget priorities in the community. Perhaps water quality is less important to decision makers than social service programs or development.

   c. If you have trouble identifying root causes, try to fill in the blanks of the following sentence for each problem identified:

   (Problem) is/are the result of *(immediate, direct cause)* which is caused by *(root cause)*.

   For the example described above, a possible answer would be:

   High fecal coliform levels (problem) are the result of *point source pollution* (immediate, direct cause) from leaky pipes which is caused by *uninspected sewer lines due to budget constraints* (root cause).

   d. Come up with a root cause of a problem even if you do not have enough information to be certain if it is truly an important factor. Later you will have the opportunity to research a problem in greater detail (Activity 9.4). The

Figure 9.5. Students in Hawaii learned that four endangered water birds were threatened in the Kawai Nui Marsh due to dredging.

purpose of this activity is to think critically, and to learn to appreciate and understand the complexity of environmental problems, so that when you move on to action taking, you can tackle the sources of a problem, rather than simply treating its symptoms.

Discussion Questions

1. Were you aware of most of the water problems that your class identified during this activity?
2. What surprised you?
3. What problems interested you the most? Why?

Activity

Visualizing the Future

(Recommended)

Objectives:

- Visualize future state of your river or stream.
- Promote written and artistic expression.
- Apply previous experiences to a new activity.

Materials:

Pencil, paper, colored pencils and paints (optional)

Time:

Approximately 40 minutes.

Background Information

Visualization is a technique used by individuals to improve their performance. For example, athletes rehearse in their minds the sequence of steps required in their sport. Visualization can also be used by people to help solve problems, and to envision solutions.

In this activity, you will develop your personal vision of a watercourse. What are your hopes for the future condition of the river? What changes would you like to see in how humans impact the river system? Draw upon prior experiences with the river or stream—mapping, interviewing people and water quality monitoring—as you visualize the future.

Procedures

1. Visualize how you would like the river to be in the future, and record your thoughts, feelings and impressions in writing and/or art. Some suggestions that may be helpful are:

   a. Clear your mind for writing. A moment of silence can help you to focus on this activity.

b. Reflect upon your experiences and knowledge of the river or stream. Think of both positives and negatives.

c. Try to be specific in writing down your thoughts. "I want the river to be clean" does not convey much information. Why do you want a clean river? So that people can swim in the river, or safely use it for drinking?

d. Discuss changes or improvements that you would like to see. Words like "reduce," "eliminate," "improve," and "increase" demonstrate movement or changes.

Figure 9.6. A Native American youth reflecting on the value of the Walpole Island marsh in Canada on the future of her people.

    e. Consider young people in the next generation, 20-30 years from now. What kind of a river would you like them to experience?

    f. You may want to assume the role of a fish, or other organism. What kind of an aquatic environment would the fish like? What are some changes that would have to occur?

2. You may wish to share your personal visions with the entire group. Student writings and art will convey goals and aspirations for the river that are shared by many students. Some ideas for achieving these goals will emerge within the writings and can serve as the groundwork for gathering further information on water quality problems, and designing an action plan.

Discussion Questions

1. What new thoughts were generated as a result of your visualization activity?

2. Do you feel what you visualized can become a reality?

3. Did this activity bring forward many new ideas?

Activity

(Recommended)

Selecting an Issue to Address

Objectives:

- Develop problem-solving skills: prioritizing problems and generating criteria for selecting one issue to address.
- Build group process skills: making choices by discussion and consensus.
- Enhance analytical skills: clearly defining and stating the problem chosen.

Materials:

List of problems and root causes generated in Activity 9.1, and blackboard or sheets of paper.

Time:

Approximately 80 minutes.

Background Information

In Activity 9.1, students generated a list of problems that affect the river and identified one or more root causes for each. If you wish to take action to solve any of these problems, you will need to spend some time selecting criteria to narrow the choices and then choose one issue to address.

Procedures

1. Generate criteria for selecting a problem to act upon. Some useful criteria are:

   a. Is the problem relevant and of high interest to students in the class?

   b. Is there adequate information about the problem?

Figure 9.7. Upon returning from a stream walk, students formed small groups to select a particular issue they would like to study.

    c. Are other people or organizations already working on the problem? (This can be very useful, since other organizations may have information and access to resources students can tap, or students might be merely duplicating efforts.)

    d. Is the problem too large or too complex for student action? If so, can it be redefined, or simplified in such a way that students can take meaningful action(s) to address the problem?

    e. What kinds of resources will you need to tackle this problem? Are these resources available? (Resources include money, time, skills, equipment.)

    f. What is the group's time frame for working on this project?

    g. What kind of action will most likely be appropriate for solving this problem? Is this level of action feasible for students?

2. Evaluate the list of problems in light of these criteria.

3. Next, agree on a problem to address.

4. Students should then work together to develop a precise statement of the problem they have selected. Make as many additional refinements to the problem definition as necessary at this point.

5. Define what you see as a successful outcome of your action(s).

6. Finally, evaluate and critique this problem statement as a whole class.

Discussion Questions

1. Was it difficult to reach consensus on your issue to be studied?

2. What influenced your decision to select the issue to be studied?

3. Do you think your class can help resolve your selected issue? Why?

Activity

(Recommended)

Contacting Organizations and Decision Makers

Objectives:

- Identify community resources.
- Gather information related to the problem selected.
- Develop phone and personal interviewing techniques.
- Learn how to write effective letters.

Materials:

Newspapers, telephone directories, government directories, reference books, periodicals, etc.

Time:

40-80 minutes (This process could potentially take longer, depending upon how much research the students need to do. Ideally, much of their work can be assigned as homework over the course of several days.)

Background Information

This activity is designed to help develop basic research skills which are essential to effective problem-solving. It is important to learn to gather information from diverse sources and to critically evaluate this information, to resolve environmental problems.

This activity can be used to help research the problem you selected in Activity 9.3. The research you conduct will provide you with a better understanding of the problem, and prepare you to develop an action plan (Activity 9.5). In classes that are not undertaking major actions, this activity can be used as an exercise to help improve research skills.

Procedures

1. You have raised many difficult questions throughout the course of the previous activities. Now you have the chance to search out some answers. In small groups or individually, generate a list of questions and issues. If your class is working on an action project, it may be appropriate to discuss these concerns with the entire group. Prioritize the information your class needs in order to better understand and work towards a resolution of their chosen problem.

2. Discuss what agencies or individuals would be appropriate to contact to find answers to your questions.

   a. Good contacts include natural resource or environmental agencies, environmental organizations, local municipal governments, industries, public works departments, etc.

   b. Useful resources for finding contacts are: phone directories, governmental directories, periodicals, newspapers, other reference books, and the local library.

3. Once you have identified contact persons or organizations, the next step is to phone, write or visit (whatever is most appropriate for the situation). Try role playing what you will say, and read your letters aloud to get feedback from others.

4. Ask contacts for other information leads. Often, people working in regulatory agencies or other governmental organizations have developed a network of contacts that students might also utilize.

5. Report the results of your inquiries to the whole class. Be certain to discuss and analyze the information that each student (or group of students) presents.

Figure 9.8. A student using the telephone directory to identify specific resources for collecting additional river information.

6. Some additional suggestions:

a. Because of their interdisciplinary nature, complex issues may require several rounds of information gathering. You will need patience and persistence!

b. Keep a record of phone numbers and addresses, when people were contacted, and the subject of the conversation or letter for later reference.

c. Consider asking one or more contact persons to make a presentation to the class. This is an excellent way to learn about an issue.

Discussion Questions

1. Were the organizations and decision makers you contacted helpful to you?
2. How could they have been more helpful?
3. What surprised you in this process?

Activity

(Recommended)

Developing an Action Plan

Objectives:

- Identify a variety of potential actions to resolve the identified problem.
- Select an appropriate and viable action (or actions) to implement.
- Outline specific steps in the implementation process.
- Build group decision-making skills.
- Develop research skills.

Materials:

Data sheets.

Time:

120 minutes.

Background Information

Once you have selected a problem and conducted some preliminary research, the next step is to develop an action plan to solve it. It is very important that the class carefully determine realistic goals and objectives for the action plan. How will you know when you have been successful? In other words, how should you define success?

Some students will have very ambitious goals for their class project. Students and teachers will need to determine what is feasible for the class to undertake. Raising awareness in classmates about a pollution issue is a viable project goal in itself. However, solving any environmental problem is challenging and time-consuming, requiring commitment and patience. But if the class has thoughtfully identified a problem and has set realistic goals for the action plan, the problem-solving experience should be very rewarding and exciting.

Procedures

1. Brainstorm possible actions to address your chosen problem. You may benefit by learning more about the various levels of action-taking available to you. Some possible actions include:

   a. Persuasion: used to try and convince others that a certain course of action is correct, or that certain behaviors need to change. Persuasion can take the form of logical reporting of facts, experiential awareness-building, or an emotional appeal. Examples include letters to the editor, presentations to classmates, parents, school boards, or city council members, posters, etc. Most strategies include some form of persuasion.

   b. Consumerism: operates on the principle that "money talks." This involves buying or not buying a product in order to influence the producers' behavior. It is usually effective only when a group of people agree to act together. Perhaps a class could try to convince parents and other students to stop buying hazardous household products by providing examples of non-hazardous alternatives.

   c. Political Action: includes any strategy that pressures political groups or government entities to take a certain action. Distributing petitions, writing letters to political figures, supporting an environmental referendum, and speaking before the city council or school board are all examples of political action.

   d. Eco-management: any physical action that improves the environment is eco-management. Campaigns to pick up trash along the riverbanks, replant eroded banks, or dislodge log-jams are typical eco-management strategies.

Figure 9.9. Small groups of students were formed to share information they had collected over the weekend.

2. Another way to help your class prepare an effective action plan is to learn from the experiences of others. Discuss and evaluate this case study, using the questions which accompany it as a guideline. Apply what you learn from this example to your own problem solving process. Look for other local examples or successful action-taking.

3. After you have generated possible strategies, develop action-taking criteria, just as you did to select an appropriate problem. Some criteria to consider include:

   a. How many steps does this strategy involve?

   b. Does this strategy involve the whole class?

   c. How will the community react to this action strategy?

   d. What level of intervention is being considered? Is the action aimed at personal changes, school level changes, or larger-scale changes in the community?

4. Select one or more action strategies based upon a consideration of these criteria. Be extremely conscious of the issue of success. If you select a strategy which you cannot possibly accomplish due to time or logistical constraints, you will

feel very discouraged. This may be the first experience with problem-solving for many students; help reduce the likelihood of failure, so that everyone becomes excited about future problem-solving opportunities.

5. After you have selected an appropriate action strategy, clearly state the problem and the strategy you have adopted to solve it once more.

6. Finally, develop a time-line and list of step-by-step procedures to aid in implementing the action plan. Also, consider what kinds of additional information might be helpful.

Discussion Questions

1. How did your group/class work together in developing an action plan?

2. What did you learn from this activity?

3. How could your group/class have been more successful?

Activity

(Recommended)

Taking Action

Objectives:

- Implement the action plan.

Materials:

All the materials used previously may be needed, depending on the problem and actions chosen.

Time:

Varies, depending upon the action plan.

Background Information

The action phase of the problem-solving process is often both an exciting and a frustrating experience. But with patience and commitment, it can be a rewarding one. The specific nature of the project will determine how best to implement the action plan.

Procedures

1. Utilize the action plan to guide the activities. Refer to the time-line and list of step-by-step procedures developed in Activity 9.5.

2. Be sensitive to the frustrations and difficulties you may experience. Be supportive and encouraging.

3. Be considerate of conflicting viewpoints. If the project goal is at all controversial, expect some people (perhaps even members of the class) to resist your efforts. Try to think broadly and empathize with a number of perspectives as you proceed. Consider the following questions:

    a. Who will welcome our project and why?

    b. Who will oppose the project and why?

    c. Who might not listen to our statements? What is our best approach with these people?

    d. How can we better understand other people's views?

4. What special planning or preparation does the project require?

    a. Will the project involve making a presentation at a local government meeting or to a board of directors? This is an excellent opportunity to experience participatory democracy and develop public speaking skills. Do role-plays or mock meetings to get ready for your presentations.

    b. Will the project require significant research off school property such as visits to sewage treatment plants, etc.? You will need to consider issues of safety, parental permission, and liability. It might be wise to visit the site before taking the students there to look for sources of potential problems.

    c. Will the project involve the use of the media? Local newspapers can be helpful to a project by increasing public awareness of an issue. However, be certain to obtain parental permission before using pictures or interviews of students in publicity efforts.

Figure 9.10. Student preparing a poster for the shopping mall to alert the community to a health problem of their river.

Discussion Questions

1. Were you successful in the actions you took?
2. How might you have been more successful?
3. Was your class interested in the taking of action?

Activity

9.7

(Recommended)

Follow-Up

Objectives:

- Evaluate the effectiveness of the action project (if appropriate).
- Evaluate the water quality monitoring program.
- Describe the problem solving process.
- Consider the applicability of skills gained from involvement in a water quality monitoring program to future attempts at problem solving.

Materials:

Paper, blackboard or poster board, pens

Time:

Approximately 40 minutes

Background Information

Evaluation of the action project and the entire water quality monitoring program (the whole series of activities they have been involved in) is a very important part of the learning process. You will benefit from a critical assessment of the impact(s) of your actions, and a sharing of thoughts and feelings about the program.

This wrap-up should serve as an affirmation of the hard but meaningful work your students have done. It will highlight some of the problems you encountered and lead to suggestions which will facilitate future attempts at problem-solving.

Procedures

1. Use the "plus-minus-change" method to evaluate the effectiveness of your action project.

    a. On the blackboard or a large piece of paper create three columns and label them "plus," "minus," and "change."

    b. Consider the question: "How well did our action address the problem we identified?" List what you liked about the action project under the "plus" column and what you did not like under "minus." The "change" column is for listing any changes you would like to make and how the action could have been improved.

    c. The teacher can join in this activity by offering feedback as well.

2. Reflect on the problem solving process you used.

    a. Each student should write down a brief:

        1) Restatement of the problem you decided to work on;

        2) Outline of the procedure you followed to address the problem;

        3) Summary of your action plan; and

        4) Description of the result of your actions.

    b. Have a discussion in which students share their depictions of the problem solving process. Use the following questions as a guide:

        1) Do students' perceptions of the problem solving process differ? If so, why? (Was there adequate communication in the class? Was everyone fully involved throughout the entire program?)

        2) What part of the problem solving process was most difficult or frustrating? Why? What can you do in the future to make it easier?

        3) How did you feel when you completed the action project? Were you successful? Why or why not?

Evaluation of the Water Quality Monitoring Program

1. Follow the same procedure for the "plus-minus-change" evaluation outlined above, but focus this time on the entire program—the whole series of activities students were involved in, including the action project.

Figure 9.11. Students carrying out an evaluation of their program—plus-minus-change.

2. Write an essay in which you share your personal thoughts and feelings about your involvement in the program. Consider some specific questions, such as:

   a. What did you like most about working with this program? Why?

   b. What did you like least? Why?

   c. What did you learn from these activities?

   d. Can you and other students your age do anything to help solve water quality problems? Why or why not?

These activities provide some of the skills needed to take effective action. Another way of learning how to take action—the process of action-taking—is through learning from others who have taken action. The following case studies offer real examples of action-taking.

Discussion Questions

1. Were you able to evaluate your action plan?
2. What skills did you acquire in the process?
3. What was the most valuable part of this program?
4. What was of least value?
5. How might this program be helpful to you in the future?

International Case Studies of Action-Taking

Case studies can be an effective way of showing how others have approached similar environmental problems. The action-taking process is made real through case studies. Although environmental problems reflect the environment and culture in which they occur, many of the approaches to solving problems are similar for related types of problems.

The environmental case studies offered here come from GREEN programs in the United States and Australia. These case studies may give hope to others struggling with environmental problems, showing that they too can make a difference.

The Rouge River Project

The Rogue River flows through Detroit, Michigan (USA) and its metropolitan area in four separate branches totaling about 200 kilometers. It is a very urbanized river system and catchment; nearly 1.5 million people live and work in this 900 square kilometer area. In the upper reaches of the four Rouge branches the land is rolling, with extensive urban sprawl, including housing developments and shopping centers. As one follows each river branch toward its confluence with the Detroit River (separating the United States and Canada), the river flows through heavily industrialized areas and skirts the City of Detroit.

There are 48 communities in the Rouge River catchment, including parts of Detroit, that contribute nonpoint source pollution from paved areas and other impervious surfaces, as well as pollution from combined sewers. Combined sewers carry both polluted stormwater runoff and sewage from homes and businesses. During rainfall and snowmelt, the sewer releases built-up pressure in the sewers by dumping untreated sewage into the river. In the Rouge catchment, there are 168 known combined sewers entering the river, contributing over 2 billion gallons of untreated sewage and polluted stormwater runoff to the river every year. The heavy loading of sewage and polluted stormwater runoff means the Rouge River poses a health risk for people exposed to the water and makes the river unfit for pollution-intolerant organisms.

Severe development in the tributary headwater streams has also led to channelization of these tributaries, loss of riparian vegetation, erosion and sedimentation, and loss of valuable aquatic habitat.

The Rouge Education Project

The Rouge Education Project began in 1987 with 16 secondary schools monitoring the Rouge River. Students and teachers measured dissolved oxygen levels, total or fecal coliform, pH, nitrates, temperature,

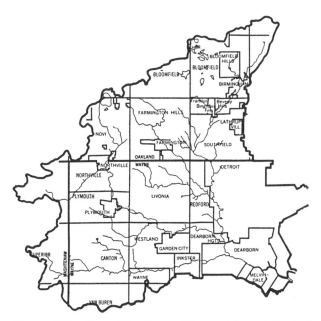

Figure 9.12. Map of the Rouge River catchment illustrating the four major branches, the main stem, and the 48 major communities.

BOD, phosphates, turbidity, and total solids at a sampling site on the river (see Chapter 8). Some students sampled and identified benthic macroinvertebrates, and others conducted stream surveys in the river and along its banks (see Chapter 7).

In 1996, the project has expanded to include 100 primary and secondary schools that conduct water quality tests, sample benthic organisms, conduct stream surveys, run simple bioassays, and measure velocity and discharge in the Rouge River catchment. Through student observations and measurements, potential problems are sometimes uncovered that lead to action-taking. The following two case studies illustrate the process of action-taking within the Rouge catchment.

Case Study One: Students Uncover A Problem

Students from a secondary school in the Detroit area participating in the Rouge Education Project measured fecal coliform at two sites along a tributary of the Rouge River near their school. The students selected two sampling locations after making a survey of their river reach and observing a suspicious-looking pipe entering the river. In order to measure any impact the pipe might be having upon water quality, the students took water quality measurements upriver and downriver of the pipe.

They conducted each of the water quality tests mentioned earlier, but stressed the fecal coliform measurement. Their testing revealed high fecal coliform levels (>2,000 colonies/100 mL) downriver from the pipe, and relatively low levels (<200 colonies/100 mL) upriver from the pipe. Students knew, through their background reading, that fecal coliform levels above 200 colonies/100 mL of water represented a greater risk of pathogenic organisms being present, and a greater risk to children in the community who might wade or splash in the river. The students' finding was made even more alarming by the death of an individual two years before who had accidentally fallen into the river in a downriver community; this death was partly attributed to a virus carried by rats and found in the river water.

The students identified the source of the problem as the pipe discharging raw sewage into this part of the river. What was the next step after problem definition? The students and their teacher asked the following questions: Who is responsible for the problem? Do we have enough information to take action? What would taking action mean in this situation? The students and teacher realized that it was not realistic to try to stop the sewage effluent themselves. They decided that the City of Farmington Hills was responsible for maintenance of the sewer system in the community, so they called the City Engineer, offering to share their data and lead him to the pipe. Initially, the students' data was greeted with skepticism. After some pressure by the students and teachers, the City of Farmington Hills decided to run fecal coliform samples themselves to see if there was agreement with the student-generated data. The two sets of measurements—the City's and the students'—matched very well.

The persistence of the students led the City to take further action and solve the problem by fixing a broken pipe that was supposed to carry sewage to the treatment plant.

Case Study Two: Restoration of a Tributary

Students and their teacher at another secondary school, Plymouth-Salem High School, followed a low-cost monitoring approach to the tributary running through the school grounds. Willow Creek is a first-order stream in the Rouge River system because it flows for nearly a half km between two schools, human pressures on the stream along this reach are severe. Students at one of the schools chose to focus on this stream for an environmental, community-based project.

The first step was to conduct a stream survey, observing the amount of vegetation along the stream, the substrate, flow, signs of erosion and degradation, and human activity. Students also compared samples of benthic macroinvertebrates taken upstream along a relatively undisturbed reach to samples taken from disturbed school sites. The other simple test conducted was temperature, along both the undisturbed and disturbed reaches.

Detroit Free Press

Section A, Page 3　　　　SECOND FRONT PAGE　　　　Sunday, May 17, 1987 •

Human wastes found in Rouge

Effluent in river despite dry period

By JOEL THURTELL
Free Press Staff Writer

When students from 16 high schools in Wayne and Oakland counties tested Rouge River water earlier this month they found enough old tires, junked cars, shopping carts and empty beer bottles to stock a town dump.

They also found the unexpected — human waste flowing into the Rouge from a sewer in Farmington Hills even though it was dry weather.

"We found that our river is as bad as we thought it would be. We found no living organisms," wrote River Rouge High School student Malika Noland. "There were oil slicks floating on top of the water, and garbage. The water was dark green and smelled very bad."

Effluent entering the river during rainy periods may be disgusting, but that's how the system was designed in the 1940s and 1950s — to let storm drains conduct sewage to the river at times of peak water flow.

But the dry weather sewage measurements by North Farmington High School students on May 7 shocked James Murray, the Washtenaw County drain commissioner who also is chairman of the State Water Resources Commission and president of Friends of the Rouge, a private group dedicated to improving water quality in the Rouge.

"It tells you the CSO (combined sewer overflow) has got crap going out of it, and that's illegal during dry water flow," Murray said.

HE DISCUSSED implications of the water test results Saturday at Detroit's Redford High School with about 50 students from the 16 Rouge basin high schools participating in the interactive Rouge River Water Quality Project sponsored by Friends of the Rouge.

But even before Saturday's meeting, students from North Farmington High wrote in the project's computer bulletin board that "we are having the city engineer of the city of Farmington Hills come to our class May 28 for us to learn of the the present system here and for us to share our data with him. Our CSO was still draining days after the last rain."

Erin English, a 16-year-old sophomore biology student from Redford Union High School, described her efforts at drawing water samples from the Rouge's upper branch: "It was really gross — we saw a couple of tires in the water. We didn't see any fish."

Data from the 16 student groups showed consistently increasing degradation along the river, said William Stapp, a University of Michigan natural resources professor who directed the project.

The Rouge runs from spindly streams merging in Oakland County to its mouth near Zug Island in River Rouge. More than 1.5 million people live in the Rouge River basin.

Sewage was well below levels considered dangerous for swimming in Novi, Birmingham and Plymouth, but fecal coliform levels increase steadily as the river flows downstream. Near the river's mouth, a student group measured fecal coliform 127 times higher than the maximum level allowed for swimming and 25 times higher than the level allowed for boating. The levels were 25,400 times high-

Figure 9.13. A writer of the local newspaper interviewed the class and carried this feature article on the students' investigation. Reprinted by permission of the *Detroit Free Press*, May 17, 1987.

Through their observations and measurements, students found the following:

➤ the stream channel was littered in places with paper, cans, and bottles from careless students;

➤ the vegetation along the school reach was denuded, especially where students crossed or jumped over the stream;

➤ the stream substrate was softer (silt-like), than upstream substrates, which tended to be gravel;

➤ water temperatures averaged 3-5 degrees Centigrade higher along the school site reach than the undisturbed, upstream site;

➤ benthic diversity was lower along the school site reach than the undisturbed site reach; and,

➤ velocity of the stream slowed along the school site reach.

Interpretation of These Observations

Students interpreted these observations in the following ways:

➤ lack of vegetation along the stream, caused by heavy human usage, led to erosion along the banks and sedimentation of the channel, resulting in the softer substrates;

➤ debris and litter in the stream channel created impounded areas along the school site, causing the stream's flow to decrease; this decrease in velocity, along with lack of shading, contributed to an increase in stream temperature; and,

➤ the softer substrate and warmer average temperature for this reach caused lower benthic diversity.

Problem Definition

The observations, measurements, and interpretations of the data led to student action-taking. First, it was important to get at the root causes of the problems. Why was there a lack of vegetation along the stream? Why did students throw cans, bottles, and paper into the creek? Through further information-gathering, students discovered that the school administration encouraged the mowing of vegetation up to the stream bank; years of mowing had discouraged any native vegetation and left only short grass. Students put together a student survey to try to find out why students threw garbage into the stream. They found that there were no garbage cans or recycling containers between the two schools, most students did not realize that this stream was a tributary of the Rouge River, and most students viewed the stream as degraded already and not a resource worth protecting.

Exploring Solutions and Taking Action

Students in this project generated some project ideas that would help restore the stream and develop a caring attitude on the part of other students towards the stream. Students took action in the following ways:

➤ A team of students and their teacher convinced the school administration and the school board to leave an unmowed vegetative buffer along the school site reach. Students then began to plant native wetland vegetation along the stream.

➤ Students found garbage cans, painted them attractively, and placed them at strategic locations between the schools.

➤ Students constructed a sign that identified the creek as part of the Rouge River system and designated a natural study area along much of the creek under restoration.

➤ Students sought approval to remove unwanted debris from the stream to speed up the flow.

➤ Students began to develop an outreach education program in which high school students would conduct nature walks for local primary school students along Willow Creek.

Case Study Three: Australian Case Study—
Mary River, Queensland

The "Waterwatch" water quality program began with one high school monitoring the headwaters of the Mary River in 1989. Presently, most schools, and multiple community groups and governmental bodies are involved in monitoring along the entire 305 km stretch of the Mary River, culminating in a student-community presentation and action oriented program.

I **Figure 9.14.** Small groups analyze the results of their investigations.

Figure 9.15. A sign was constructed and posted along the intersection of the river and a road noting the groups that cooperated on the project.

Based on the water quality monitoring data and findings, there has been a growing community interest in the health conditions of the Mary River system.

The river monitoring activities led to the formation in late 1993 of the Mary River Catchment Coordinating Committee, an initiative of the Queensland Department of Primary Industries. This committee is composed of representatives from citizen and governmental groups and student representatives from throughout the catchment. Its task is to maintain and improve the health of the Mary River and its catchment, by co-operative, community oriented river activities.

One issue that concerned the communities along the river was stream bank erosion. At Conondale, a farming community in the upper catchment, residents were concerned that they could lose their bridge across the river because of stream bank erosion. Based on this concern, local high school students and community-based groups drew up a plan to help stabilize the Mary River bank in the area near the bridge. Participating groups became involved in a variety of complementary land-use tasks.

The local Bush Fire Brigade started the process by removing the non-native weed growth on the site. The local Council provided a bulldozer to shape the bank back to a 45° slope. A Queensland Forest Service extension officer provided advice on suitable endemic flora and planting techniques and helped to formulate a restoration plan. Teachers and students from Conondale primary school and community residents planted the bank with millet to help stabilize the soil. High school students planted native trees on the riparian land and bank, under the supervision of members of the Barung Landcare group. Local farmers provided mulch. The Bush Fire Brigade became involved in follow-up watering.

Given that the site is next to the bridge on the main road winding down the river valley, this visible demonstration site has served as a living model for other stream improvement projects in critical areas along the Mary River. These projects are supported by the Mary River Catchment Coordinating Committee and Mary River Waterwatch.

Appendix A

Physical-Chemical Testing Procedures; Safety Guidelines; Ordering Information; Construction of Sampling Equipment

Physical-Chemical Testing Procedures

Procedures are offered here for each of the nine water quality measures that constitute the Water Quality Index. Background information about each of these tests is included in Chapter 8. For dissolved oxygen, nitrates, and total phosphates the procedures differ from those included in commercial kits. Those who are able to purchase commercial kits, or who have the hardware from commercial kits and are trying to replace chemicals, have access to procedures for those kits. The procedures for dissolved oxygen, nitrates, and phosphates are specifically for option 3 users; investigators who must find chemicals and hardware. Some measurements, like temperature and total solids have only one option.

▼

TEST PROCEDURE #1
Dissolved Oxygen

The following procedure is known as the azide modification of the winkler method for the dissolved oxygen test. This is an appropriate method for most waters of the world because it is used to test for sewage, effluent, and stream samples.

The third option is to manufacture the reagents using chemicals that are available from a chemistry lab. These chemicals may be difficult for students to obtain and mix, but a chemistry teacher would certainly have the expertise to combine chemicals to make the reagents. A scale that measures to 1/1000 of a gram is necessary. In addition, the chemicals must be mixed in quantities that are far beyond the needs of a single water quality testing group. The reagents are usually mixed to volumes exceeding 100 mL, but only a fraction of a milliliter is needed to perform each test. To effectively utilize laboratory time and resources and prevent waste (some of the chemicals have only a two week shelf life), networking among testing groups will facilitate the use of these chemicals during a brief period of time.

Reagents

- *Manganous sulfate solution:* Dissolve 364 g $MnSO_4•H_2O$ in distilled water, filter, and dilute to 1 L.
- *Alkaline iodide-azide solution:* Dissolve 500 g NaOH and 135 g NaI in distilled water and dilute to 1L. To this solution, add 10 g NaN_3 dissolved in 40 mL distilled water. If Na+ cannot be found KOH and KI would be fine.
- *Sulfuric acid:* concentrated, one mL is equivalent to about 3 mL alkaline iodide-azide reagent.
- *Starch solution:* Dissolve 2 g soluble starch and 0.2 g salicylic acid in 100 mL hot distilled water.
- *Sodium thiosulfate titrant:* Dissolve 6.205 g $Na_2S2O_3•5H_2O$ in distilled water. Add 1.5 mL 6N NaOH or 0.4 g solid NaOH and dilute to 1000 mL.

Hardware

- 200 to 300 mL
- Bottle with ground glass stopper
- Pipets for adding reagents to sample
- Dropper, or syringe, for the titration steps. (Need to calibrate dropper so that a defined number of drops is = to mL of titrant.)

Procedures

1. To the sample collected in a 250 to 300 mL bottle, add 1 mL $MnSO_4$ solution.
2. To this add 1 mL alkali iodide-azide reagent.
3. Stopper carefully, to avoid entrapping air bubbles and mix by inverting the stoppered bottle several times.
4. A precipitate or floc will form; allow this floc to settle to about half the bottle volume.
5. When floc has settled, add 1 mL conc. H_2SO_4.
6. Restopper the bottle and mix by inverting several times until all of the floc has dissolved into the sample.
7. Titrate with the 0.0021 M $Na_2S_2O_3$ solution to get a pale straw color.
8. Add a few drops of the starch solution, which will turn the sample blue.

9. Continue titration with $Na_2S_2O_3$ until the blue color turns clear.

10. 1 mL of 0.0021 M $Na_2S_2O_3$ = 1 mg dissolved oxygen/L

Taken from: *Standard Methods for the Examination of Water and Wastewater,* 16th ed. American Public Health Association, Inc., New York.

▼
TEST PROCEDURE #2
Fecal Coliform

Reagents

Fecal coliform m-fc agar (directions for a liter):

- Agar 15.0 g
- Lactose 12.5 g
- Tryptose 10.0 g
- NaCl 5.0 g
- Protease Peptone #3 5.0 g
- Yeast Extract 3.0 g
- Bile Salts 1.5 g
- Aniline Blue 0.1 g
- Rosolic Acid solution 10.0 mL (To make rosolic acid, measure 1.0 g rosolic acid and add to 0.2 N NaOH and bring volume to 100 mL.)

To make agar: Add 10 mL rosolic acid solution to 950 mL deionized water and add all of the other components. Bring volume to 1.0 L and gently heat to boil. Pour into sterile petri dishes.

Alcohol

Hardware

- Filtration system
- Petri dishes w/absorbent pads
- Pipet
- Matches
- Filters

- Plastic bags
- Forceps or tweezers

(See Figure 8.7, Chapter 8 for a drawing of the fecal coliform equipment.)

Option One Procedure

Regardless of the method utilized for isolating the bacteria onto the filter paper, the same procedure should be followed for collecting the sample. Procedures for measuring fecal coliform levels will vary with each option. Option one below includes commercially available equipment and materials, and option two utilizes Tupperware™ and homemade chemicals.

Collecting Sample

1. Remove the stopper or cap of the sampling bottle just before sampling and avoid touching the inside of the cap.
2. If sampling by hand, using gloves, hold the bottle near its base, plunge it (opening downward) below the water surface, turn the bottle underwater into the current and away from you.
3. Avoid sampling the water surface because the surface film often contains greater numbers of fecal coliform bacteria than is representative of the river. Avoid sampling the bottom sediments for the same reason, unless this is intended.
4. The same general sampling process applies when using the extended rod sampler.
5. When collecting samples, leave some space (an inch or so) to allow mixing of the sample before pipetting.

It is a good idea to collect several samples from a particular river location to minimize the variability that comes with sampling for bacteria. If possible, sanitation should occur between sampling sites. Ideally, all samples should be tested within one hour of collection. If this is not possible, the sample bottles should be placed in ice and tested within six hours.

Procedures

6. Sanitize forceps: To sanitize forceps, dip in alcohol and burn alcohol off with a flame (an alcohol lamp works well). Do not place the hot forceps back into the alcohol.

A word about sterilization . . .

It is essential to sterilize sample bottles, pipettes, and filtration system before sampling. Sterilization can be accomplished by using an autoclave, 121° C for 15 minutes. If an autoclave is not available, the home economics department may have a pressure cooker that they might be willing to lend to the water quality monitoring program. If a pressure cooker is used, please be sure that it has a working pressure gauge. The gauge may be checked with the county cooperative extension service. The pressure cooker should be run at 15 psi. to properly sterilize sample bottles, pipettes and filtration system.

If these two pieces of equipment are unavailable, an oven can be used. The oven must attain a temperature of 170° (± 10° C) for not less than 60 minutes. The plastic filtration system cannot, however, be placed in a dry oven because the system will melt. The same holds true for plastic sampling bottles. The filtration system can, however, be placed in boiling water for 5 minutes to sanitize it. Petri dishes, culture media, absorbent pads, and filters are presterilized and packaged. Equipment that has been inadequately sterilized may interfere with fecal coliform growth.

. . . and sampling design

If the purpose of sampling is to determine fecal coliform levels at a river reach, then samples should be taken beneath the water surface and in the current (if there is one). If the purpose of sampling is to confirm suspected sources of fecal coliform contamination, then samples should be taken just downriver from the source (like the mouth of a storm drain), and other samples should be taken upriver from the source for comparison.

There is also wet-weather sampling and dry-weather sampling. Wet-weather sampling involves sampling during and just after a rainstorm, often at timed intervals. It is done if fecal coliform contamination is suspected from storm drains carrying urban stormwater runoff. Wet-weather samples can then be compared to samples taken during a period of dry weather (dry-weather samples). The bottles used for the dissolved oxygen test might also be used for the fecal coliform test.

Try to avoid sampling stagnant areas of rivers. The extended rod sampler is an effective device for obtaining a sample in the current. If sampling rivers in which little current exists, push the sample bottle underwater away from your body, thereby creating a current.

7. Using the sanitized forceps, place an absorbent pad in the pre sterilized petri dish. Be careful not to touch the pad or the inside of the petri plate or lid with your fingers.

8. Add broth to the petri dish. The broth is liquid food for fecal coliform bacteria. Put the top on the petri dish, and set aside.

9. Sanitize forceps with alcohol and flame again.

10. Uncouple the top half of the filtration system and place the sterile filter paper grid side up on top of the filtration system's holed area with forceps. Be sure the filter paper is completely flat with no wrinkles.

11. Recouple the top half of the filtration system to the bottom half.

12. Before taking a sample, pour a small amount of distilled water into the filtration system with the top cover off.

13. Place the pipette into the water to be sampled. The ideal number of fecal coliform colonies is 20-60 on a petri plate. The range (20-60) reduces the effects of colony crowding on sheen or color development.

Suggested fecal coliform sample volumes (in milliliters)					
Drinking water	100.0				
River water	5.0	2.0	1.0	0.1	
Stormwater runoff			1.0	0.1	0.01
Raw Sewage, CSO's		1.0	0.1	0.01	0.001

Adapted from *Standard Methods for the Examination of Water and Wastewater.*

14. Place the end of the pipette into the open hole on the half of the vacuum assembly containing the filter paper. Release the water sample into the unit.

15. With the filtration system level, use the syringe to create a vacuum, drawing all the sample and distilled water through the filter while swirling. Swirling reduces the number of bacteria adhering to the upper filtration system. Remove the plunger from the vacuum assembly when pushing the plunger in so as not to push air back into the filtration unit, which would force the filter off the absorbent pad. Use the syringe to draw water through the filter until it appears dry.

16. Uncouple the top half of the assembly, and carefully remove the filter paper with the sanitized forceps.

17. Open the top of the petri dish, and slide the filter across and into the dish, with the grid side up. Petri dishes should be incubated within 30 minutes of filtering the sample; this will ensure heat shock of any non-fecal coliform organisms. Be sure to record the date, site, and volume of sample on the frosted part of the petri dish.

18. Enclose the petri dish in a waterproof bag (to avoid leakage) and then put into the water bath. Dishes may also be sealed with waterproof tape (freezer tape) to avoid leakage. For fecal coliform, incubate for 24 ±2 hours at 44.5°C. (Temperature must be maintained within a range of ±0.25°C of 44.5°C). Petri dishes should be inverted (side with filter paper and absorbent pad up) to avoid condensation. Total coliform can be incubated in a warm room. Please wash your hands after this test.

19. After incubation, carefully count the bacterial colonies on the filter, using a magnifying glass (10x) or unaided eye. You might want several people to verify the bacterial count. Each bluish spot is counted as one fecal coliform colony. Fecal coliform colonies should be examined within 20 minutes to avoid color changes that occur with time.

▼

TEST PROCEDURE #3

pH

Option one for the pH test is described in Chapter 8 and involves the use of a pH meter. Option two involves the use of a kit and option three is described below. Test papers are by far the most inexpensive way to test for pH in water. Using these papers requires no extra training or equipment. They are extremely inexpensive (US $4 for 200-300 tests) and can be made even less expensive simply by cutting them in half lengthwise. Accuracy, however varies considerably on test strips. Some have errors as high as ±10 percent. Check with your local distributors of pH papers for accuracy. Information about suppliers of pH test papers can be found on pp. 267-268.

Procedures

1. Follow the sample collection procedure outlined above.

2. Dip the pH paper into the water sample for about 15 seconds or until a color change occurs.

3. Compare the color from the dip stick to the comparison sheet included with the pH papers.

4. Record your results.

▼

TEST PROCEDURE #4

Biochemical Oxygen Demand

Reagents and Hardware

(see dissolved oxygen test)

Procedures

1. Fill two glass-stoppered dissolved oxygen (DO) bottles (one clear and one wrapped in black tape) with sample water, holding them for two to three minutes between the surface and the river bottom. If sampling by hand, remember to use gloves.

2. Prepare the clear sample bottle according to the directions for the dissolved oxygen test. Determine the DO value for this sample in mg/L. Record on the data sheets.

3. Place the black sample bottle in the dark, and incubate for five days at 68°F (20°C). This is very close to room temperature in many buildings. In tropical areas, it may be necessary to leave the bottle in a shady, moist area where evaporation will keep the bottle cool. If no incubator is available, place the blackened sample bottle in a "light-tight" drawer or cabinet.

4. After five days, determine the level of dissolved oxygen in mg/L of this sample by repeating either Option One or Two of the DO testing procedure.

5. Determine the BOD level by subtracting the DO level found in Step 8 from the DO level found in the original sample taken five days previously. Record on data sheets.

   BOD = mg/L Dissolved Oxygen (original sample) minus the mg/L Dissolved Oxygen (after incubation)

▼

TEST PROCEDURE #5

Temperature

Procedures

1. The temperature test compares the difference in water temperature between two different stream sites. Select two stream sites to test for which the physical conditions, current speed, amount of sunlight reaching the water, and the depth of the stream are as similar as possible. One of the sites should be the monitoring site in which you are running all of your tests. The second site should be about two kilometers upstream.

2. To reduce errors, the same thermometer should be used to measure the water temperature at both sites.

3. Rubber gloves should be worn if there is any chance that hands may contact the stream.

4. Lower the thermometer four inches below the surface at the river site to be tested for water quality.

5. Keep the thermometer in the water until a constant reading is obtained (approximately two minutes).

6. Record your measurement in degrees Celsius on the data sheet.

7. Repeat the test as soon as possible approximately 1.2 km upstream.

8. Subtract the upstream temperature from the temperature downstream using the following equation:

temperature downstream minus the temperature upstream = temperature difference

9. Record the difference in temperature.

▼
TEST PROCEDURE #6
Total Phosphates

The procedure below includes a digestion step (persulfate digestion) followed by the ascorbic acid method for the orthophosphate measurement.

Reagents

- *Sulfuric acid solution:* Add 300 mL conc H_2SO_4 to approximately 600 mL distilled water and dilute to 1 L with distilled water.
- *Ammonium persulfate:* $(NH_4)_2S_2O_3$ solid
- *Sodium hydroxide:* NaOH, 1N

Hardware

- Hot plate or stove
- Erlenmeyer flask (125 mL) or glassware

Procedures

Persulfate Digestion

1. Add 1 mL of H^2SO^4 solution to a 50 mL sample in a 125 mL Erlenmeyer flask.
2. Add 0.4 g of ammonium persulfate.
3. Boil gently on a hot plate or stove for 30-40 minutes, or until a volume of 10 mL is reached. Do not allow sample to dry.
4. Cool and dilute the sample to about 30 mL and adjust the pH to 7.0 with 1N NaOH. Dilute to 50 mL.

Orthophosphates

5. Add 8.0 mL of combined reagent (see below) to sample and mix thoroughly. After 10 minutes, but no longer than thirty minutes, compare the color to a color wheel.
6. Read as Phosphorus, mg/L.

Reagents

- *Sulfuric acid solution:* 5N: Dilute 70 mL of conc. H_2SO_4 with distilled water to 500 mL.

- *Potassium antimonyl tartrate solution:* Weigh 1.3715 g $K(SbO)C_4H_4O_6 \bullet 1/2H_2O$, dissolve in 400 mL distilled water in 500 mL volumetric flask, dilute to volume. Store in a cool place in a dark, stoppered bottle.

- *Ammonium molybdate solution:* Dissolve 20 g $(NH_4)_6Mo_7O_{24} \bullet 4H_2O$ in 500 mL of distilled water.

- *Ascorbic Acid, 0.1 M:* Dissolve 1.76 g of ascorbic acid in 100 mL distilled water.

- *Combined reagent:* Mix the reagents above in the following proportions for 100 mL of the mixed reagent: 50 mL of 5N H_2SO_4, 5 mL of antimony potassium tartrate solution, 15 mL of ammonium molybdate solution, and 30 mL of ascorbic acid solution. *MIX AFTER ADDITION OF EACH REAGENT. ALL REAGENTS MUST REACH ROOM TEMPERATURE BEFORE THEY ARE MIXED, AND MUST BE MIXED IN THE ORDER GIVEN.*

▼

TEST PROCEDURE #7

Nitrates

Nitrates are particularly difficult to measure because of the complexity of the procedures, and the somewhat narrow range of concentrations in which these procedures work. The Hach kits use the cadmium reduction method for determination of NO_3 levels within 0.1 mg/L and 1.0 mg/L. While this procedure is applicable for many waters, the chromotropic acid method is recommended for 0.1 mg/L to 5 mg/L.

Chromotropic Acid Method

Reagents

- *double distilled water or demineralized water*

- *stock nitrate solution:* Dissolve 7.218 g KNO_3 in distilled water and dilute to 1000 mL. (1.00 mL = 1.00 mg NO_3-N)

- *standard nitrate solution:* Dilute 10 mL of nitrate stock solution to 1000 mL with distilled water. (1.0 mL = 0.01 mg NO_3-N)

- *sulfite-urea reagent:* Dissolve 5 g urea and 4 g anhydrous Na_2SO_3 in water and dilute to 100 mL.

- *antimony reagent:* Heat 500 mg Sb metal in 80 mL conc. H_2SO_4 until all the metal has dissolved. Cool and carefully add to 20 mL iced water.
- *chromotropic acid reagent:* Purify chromotropic acid (4,5 dihydroxy 2, 7-napthalene disulfonic acid) as follows: Boil 125 mL water in a beaker and gradually add 15 g 4,5 dihydroxy 2, 7-naphalene disulfonic acid disodium salt with constant stirring. Add 5 g activated charcoal. Boil for about 10 minutes. Add water lost to evaporation. Filter hot solution through cotton wool. Add 5 g activated charcoal to filtrate and boil for 10 more minutes. Filter through cotton wool and then through filter paper. Cool and slowly add 10 mL conc. H_2SO_4. Boil solution until about 100 mL are left in beaker. Let stand overnight. Wash chromotropic acid crystals thoroughly with 95 percent ethyl alcohol until crystals are white. Dry crystals at 80°C. Dissolve 100 mg chromotropic acid in 100 mL conc. H_2SO_4 and store in a brown bottle.
- *sulfuric acid, concentrated*

Procedures

1. Prepare NO_3 standards in the range 0.10 to 5.0 mg N/L by diluting 0, 1.0, 5.0, 10, 25, 40, and 50 mL standard NO_2 solution to 100 mL with water.
2. Pipet 2.0 mL portions of standards, samples, and a water blank into dry 10 mL volumetric flasks. Use dilutions of standards and samples in the range 0.1 to 5.0 mg NO_2-N/L.
3. To each flask add 1 drop sulfite-urea reagent.
4. Place flasks in a tray of cold water and add 2 mL Sb reagent. Swirl flasks during addition of each reagent.
5. After about 4 min in the bath, add 1 mL chromotropic acid reagent, swirl, and let stand in cooling bath for 3 min.
6. Add conc. H_2SO_4 to bring volume in flasks near the 10 mL mark.
7. Stopper flasks and mix by inverting each flask four times.
8. Let stand for 45 minutes at room temperature and adjust volume to 10 mL with conc. H_2SO_4.
9. Read absorbence at 410 nm with a spectrophotometer, or compare to a color chart.
10. mg NO_3-N/L = $\dfrac{\text{ug } NO_3\text{-N (in 10 mL final volume)}}{\text{mL sample}}$

Using Test Strips
Procedures

1. Dip the nitrate test strip into the sample for one second. Swirling lightly to remove air bubbles.
2 Wait for at least one minute, but not more than five, and compare the nitrate strip with the color code. Record the results in milligrams per liter.

▼

TEST PROCEDURE #8
Turbidity

Turbidity Tube

A turbidity tube is easier to use and more precise than the Secchi Disk, although the "tube" may be somewhat difficult to find and slightly more expensive. The turbidity tube consists of a transparent, calibrated tube with a small black "cross" on the bottom.

Procedures

1. Collect a small bucketful of stream water to sample, being careful not to disturb the sediments on the stream bottom.
2. Shake the sample vigorously before examination.
3. Hold the tube over a white surface such as a sheet of paper out of direct sunlight. Gradually pour the water sample into the tube while looking vertically down into the tube.
4. Stop pouring when the black cross on the bottom of the tube is just visible.
5. Note the reading from the scale on the side of the tube.
6. Pour a small amount of the water sample out of the tube and repeat steps and 4 or have another person perform the procedure to verify the result.
7. Record this reading.
8. If the reading is above 200, dilute the sample 1:1 with distilled water (for example, if you have a 150 mL of sample, add another 150 mL of distilled water). Multiply the result by 2 to give the turbidity. If the result is still above 200, repeat the 1:1 dilution and multiply the reading by 4.

Secchi Disk

The Secchi disk consists of a eight-inch round disk divided at the axis into four quadrants, two white and two black. (See photo in Figure 1.4, Chapter 1.) At the center of the disk is an eyebolt, attached to a rope used to lower the unit into the water. A number of problems may occur using the Secchi disk in rivers. Rivers are often so sediment-laden that the disk immediately disappears upon immersion, making the measurement for turbidity "very turbid," but impossible to quantify. Or, if the river is too shallow and one can see the bottom, the Secchi disk cannot be lowered deep enough to measure turbidity. Strong currents can angle the Secchi disk downstream, making the reading inaccurate. Attempting to correct this with heavy weights can make the disk difficult to raise and lower. Safety is another consideration with the Secchi disk, as it involves leaning over the water. Instructions for constructing the Secchi disk are found in Appendix A.

Procedures

When using the Secchi disk, it is important to make sure the disk travels vertically through the water. If the disk is moved too much by the current, testing should be moved to a spot where the current is less strong, or weights could be added to the unit.

1. Lower the Secchi disk from a bridge, boat or dock into the water until it disappears. Note the number of meters/centimeters on the chain or rope.
2. Drop the disk even further, and then raise it until you can see the disk again. Note the number of meters/centimeters on the chain.
3. Add the results of step 1 and 2 and divide by two. This is the turbidity level.

TEST PROCEDURE #9

Total Solids

Hardware

- 300 mL beaker (or any large beaker)
- Oven pads or tongs

Procedures

1. Fill a bottle with at least 100 mL of the water to be sampled. Remove any large floating particles or submerged masses from the sample.

2. In the laboratory, clean a 300 mL beaker (any 300 mL bottle may be used, but a beaker provides greater surface area) and dry in a 103°C oven for one hour. The beaker may also be placed over a burner with low red heat.

3. Remove beaker from heat with tongs and allow it to cool, then weigh with a sensitive balance (to the nearest .0001 gram). Record on data sheet. It is important to wait for the beaker to cool, as thermal currents rise a hot beaker will cause the reading to be inaccurate. *DO NOT TOUCH THE BEAKER WITH BARE HANDS* because body moisture and oils will be transferred to it, changing the weight of the beaker. Use tongs, if available, or pads or gloves.

4. Transfer a 100 mL sample into the 300 mL beaker. If sample has been sitting, swirl the sample water by pipette or graduated cylinder before measuring.

5. Evaporate the sample by drying the beaker and the resulting residue in a 103°C oven overnight. Do not allow the sample to boil or otherwise allow any of the sample to splash out. Allow the beaker to cool, then re-weigh it. Record on data sheet. As before, it is important not to touch the beaker with your hands.

6. Subtract the original dry weight of the beaker from the weight of the beaker with the residue to obtain the weight of the residue (in grams).

7. Use the formula below to calculate total solids.

The formula for determining total solids is:

$$\frac{\text{Weight of residue (gm)}}{\text{Volume of sample (ml)}} \times \frac{1000 \text{ mg}}{1 \text{ gram}} \times \frac{1000 \text{ ml.}}{1 \text{ liter}} = \text{mg/l}$$

EXAMPLE: Weight of beaker and residue = 48.2982 grams

Weight of beaker   48.2540 grams
Weight of residue   .0442 grams

$$\frac{.0442 \text{ grams}}{100 \text{ ml}} \times \frac{1.000 \text{ mg.}}{1 \text{ gram}} \times \frac{1.000 \text{ ml.}}{1 \text{ liter}} = 442 \text{ mg./l}$$

Safety Guidelines for Catchment Monitoring (Chemically-Physically-Biologically)

It is vital to give proper consideration to safety guidelines prior to catchment monitoring, during catchment monitoring, and following catchment monitoring. The top priority in a catchment monitoring program must be participant safety.

The following are some appropriate guidelines to follow so that everyone is properly protected while monitoring the catchment, river system, and riparian area.

1. Safety guidelines to consider prior to monitoring:

   a. Make certain that the kits and chemicals are properly stored and labeled in your school building.

   b. Know the school policy and safety guidelines related to a catchment monitoring program, such as travel permission and health care requirements.

   c. Read the "chemical manifest" sheets enclosed with the chemical kits to know what chemicals will be involved during the testing and how to dispose of the chemicals properly.

   d. Make certain you have the safety equipment to carry out a monitoring program—rubber gloves, goggles, eye wash, hazardous waste container, waterproof boots and appropriate clothing.

   e. Reproduce this handout for students, and go over safety instructions in a thorough manner with students prior to monitoring and make certain that all instructions are clearly understood (review this handout with students that were absent during the day the safety guidelines were discussed).

   f. Visit the monitoring site for safety precautions prior to sampling, and avoid sampling from bridges that are traveled heavily by cars and trucks or shorelines with steep, muddy or slippery banks.

   g. When monitoring close to a waste water treatment plant or hazardous site, surgical masks should be worn to protect against aerosols (aerosols are windborne contaminants that can be breathed into the lungs).

   h. Contact your nearest Health Department, Department of Natural Resources, or local planning agencies for specific warnings regarding local rivers. Some stretches of river may contain dangerous water level shifts or contaminants in the sediments that would be of particular concern in benthic macroinvertebrate sampling.

   i. In general, the following items will help to ensure a safe monitoring experience:

- Safety goggles for each student
- Rubber gloves to cover the hands
- Waterproof boots to protect feet from contaminants
- Adequate clothing to protect body in inclement weather
- Pail of clean water for washing hands
- Eye-wash bottle
- Container for holding wastes
- Completely equipped first aid kit
- Access to a nearby phone in case of emergency

2. Safety guidelines to consider during monitoring:

   a. Review all rules and procedures at the site prior to monitoring.

   b. Know the level of fecal coliform in the river prior to taking the class to the river. If the fecal coliform level is higher than 200 fecal coliform colonies/100 mL of water, no person should have any part of her/his body in direct contact with the water.

   c. Always monitor with a partner and never in water above your knees or with a strong current.

   d. Take responsibility for yourself regarding safe chemical-physical-biological monitoring of the river by knowing your responsibility and doing everything according to instructions.

   e. Take the opportunity to discuss safe monitoring procedures with any student that is testing the water in a manner that is unsafe for one's health or that of any other person.

   f. Do not put fingers in or near one's eyes or mouth while monitoring due to the potential damage of chemicals to human health.

    g. Have safety goggles available for students that will be testing the water, and make certain that they are worn while monitoring. (Remember contact lenses are not a substitute for goggles.)

    h. Have gloves available for all students that will be monitoring, and make certain that they are worn while monitoring the river.

    i. Treat all fecal coliform cultures as potentially pathogenic, and therefore use gloves and forceps to avoid possible contamination. Sterilize all containers and forceps after use.

    j. Wash water should be available on the monitoring site for anyone that might need a source of water to rinse chemicals from one's eyes or body.

    k. Have marked containers for each chemical waste for proper disposal. Each disposal chemical should be kept separate and not mixed with other chemicals, so that each chemical can be disposed of properly.

    l. Report to the teacher/leader any safety violations—not using gloves, not wearing goggles while mixing chemicals, not disposing of chemicals properly, etc.

3. Safety guidelines to consider following monitoring

    a. Dispose chemicals in a proper manner—place in a hazardous waste container, neutralize according to proper procedures, take the chemicals to a disposal site.

    b. Sterilize all equipment and gloves following the river monitoring program and store properly.

    c. Act on the data collected—send the data to regulatory or planning agencies, work with supportive agencies to post waters of poor quality, prepare articles on the data collected, offer to talk on local TV or radio regarding class findings, etc.

    d. Label and store all kits and chemicals in a safe and proper place (also label all other collected data, water samples and kits).

Water Quality Testing Equipment: Range, Accuracy, Price, and Ordering Information

Range, Accuracy and Price

MANUFACTURER	INSTRUMENT	RANGE	ACCURACY	PRICE
pH				
Hach	pH Kit (1470-11)	5.5-8.5 pH units	0.1 pH	$47.50 (US)
Merck	9533 (plastic strips)	5.0-10.0 pH units	0.5 pH	
LaMotte	DC 1600	5.0-9.5 pH units	0.1 pH	$39.00 (US)
Dissolved Oxygen				
Hach	Test Kit (1469-00)	NA	0.5 mg/L	$44.00 (US)
LaMotte	Test Kit (7414)	NA	0.5 mg/L	$29.60 (US)
Nitrogen				
Hach	Test Kit (14161-00)	0.0-1.0 mg/L	0.1 mg/L	$38.25 (US)
Hach	Test Kit (1468-03)	0.0-50.0 mg/L	1 mg/L	
Tintometer	Nitrate Kit	0.1-1.0 mg/L		
LaMotte	Nitrate (3354)	0.0-15.0 mg/L		$35.00 (US)
Ohmicron	Nitrate	0.0-15.0 mg/L		
Phosphorus				
Hach	Test Kit (2250-01)	0.0-1.0 mg/L	0.2 mg/L	$109.75 (US)
		0.0-5.0 mg/L	1.0 mg/L	
Tintometer	Test Kit	0.0-4.0 mg/L		
LaMotte	Test Kit (7884)			$78.95 (US)
Turbidity				
LaMotte	Kit (7519)	5-200 JTU	5 JTU	$33.75 (US)

Fecal Coliform Supplies

	HACH		MILLIPORE	
Filter Holder/Sterifil System	22544-00	$63.50/ea.	XX1104700	$62.00 (US)
Filters/type HC	13530-01	$85.00/200 pkg.	HCWG0473	$80.00 (US)
Petri dishes w/pads	14717-99	$27.95 for 100	PD10047SO	$24.00 (US)
Syringe/hand vacuum pump	23239-00	$10.95/ea.	XKEM00107	$35.00 (US)
Fecal coliform media	14104-15	$16.00/pkg. of 15	M00000P2F	
Pkg. of 50	$44.00 (US)			
Total coliform media			M0000002E	$20.00 (US)

Ordering Information

Hach Chemical Company
World Headquarters
P.O. Box 389
Loveland, CO 80537
Tel: 1-800-227-4224 for calls from
within the United States
1-303-669-3050 for calls from Canada,
Latin America, the Carribean, the
Far East, and Pacific Basin. You may
call collect.
FAX: 303-669-2932 Telex: 160840

Hach Europe
(serving Europe, Africa, the Middle East,
Near East and North Africa)
Chaussée de Namur 1
B-5150 Floriffoux
Namur Belgium
Tel: (32)(81) 44.53.81
FAX: (32)(81) 44.13.00 Telex: 846-59027

Millipore
P.O. Box 255
Bedford, MA 01730
Tel: 1-800-645-5476 (East Coast and Canada);
1-800-632-2708 (West Coast)
FAX: 1-508-624-8873 (East Coast)
1-415-952-1740 (West Coast)

LaMotte Company
P.O. Box 329
Chestertown, MD 21620
Tel: 1-800-344-3100 410-778-3100
FAX: 410-778-6394

Tintometer
309A McLaws
Williamsburg, VA (USA)
(804) 220-2902
Other offices in Salisbury, England (UK)
and Dortmund (Germany)

Millipore Worldwide (telephone numbers)
Australia: 008-222-111
Austria, Central Europe, C.I.S., Africa,
Middle East and Gulf: 43-877-8926 (Austria)
Baltic Republics: 90-804-51-10
Belgium and Luxembourg: 02-242.17.40
Brazil: 011-548-7011
China: 86-1-5135114/5135116-7 (Beijing)
Czech Republic: 42-2-35-02-27/2-35-23-75
Denmark: 46-59-00-23
Finland: 90-804-51-10
France: 1-30.12.70.00
Germany: 06196-494-0
Hong Kong: 852-803-9111
Hungary: 36-1-166-86-74
India: 91-808-394657/396320
Italy: 02-25078.1
Japan: 03-3474-9111
Korea: 82-2-5548305
Malaysia: 03-757-1322
Mexico: 525-576-96-88
The Netherlands: 01608-22000
Norway: 22-67-82-53
Poland: 48-2-669-12-25
Puerto Rico: 809-747-8444
Singapore: 65-253-2733
Spain: 91-729-03-00
Sweden: 08-628-69-60
Switzerland: 01-945-3242
Taiwan: 886-2-7001742
U.K. and Ireland: 0923-816375
all other countries: 508-624-8622 (U.S.A.)

How to Obtain Aerial Photos and Satellite Images

Sources of aerial photography can be obtained by requesting the pamphlet *How to Obtain Aerial Photographs available from United States Geological Survey (USGS)*, User Services Section, EROS Data Center, Sioux Falls, South Dakota 57198. The telephone number is 605-594-6151.

Landsat images are available for the 50 states and for most of the earth's land surface outside North America. A useful brochure, *Landsat Products and Services*, is also available through EROS.

Your local catchment council, planning bodies, universities, or environmental agencies might have aerial photographs or satellite images for classroom use. Aerial photographs and satellite images can be photographed on slides (with permission), and reproduced in color into 11 x 17 prints on some copy machines.

Construction of Simple Sampling and Monitoring Equipment

Homemade sampling devices are often much less expensive than purchased devices. The directions for the construction of simple sampling devices and equipment comes from the Tennessee Valley Authority (TVA) and is used in this manual with their permission.

▼

EQUIPMENT CONSTRUCTION PROJECT #1

Attached Algae Sampler

This sampler is used as an artificial substrate to collect attached algae. Place microscopic slides in relatively clear, shallow waters. In waters of heavy silt, orient the slides vertically. Artificial substrates are sometimes moved, stolen, or vandalized, so place in a seldom used area of the stream. See activity 7.13 in Chapter 7.

Materials

- 4 microscope slides
- 1 brick
- waterproof adhesive (handy-tac)
- fishing line
- bobber
- scissors

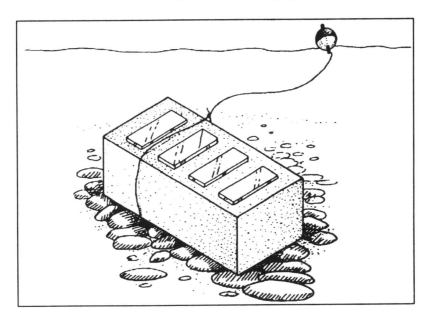

Directions

1. Take waterproof adhesive and tear off eight small pieces and roll into small balls.

2. Attach microscope slide to brick by placing two pieces of adhesive on brick spaced so there will be one for each end of the slide. Firmly press a clean, dry slide onto two pieces of handy-tac until it is stuck tightly to the brick. Repeat with remaining slides.

3. To attach marker, tie it securely to brick as shown.

▼

EQUIPMENT CONSTRUCTION PROJECT #2
Multi-Plate Sampler (Hester-Dendy Sampler)

The multi-plate sampler is an artificial substrate for colonizing benthic macroinvertebrates. This sampler is particularly useful in collecting benthic macroinvertebrate communities where the kick screen or D-frame net cannot be used effectively (e.g., deeper waters, dangerous waters because of current, low visibility, etc.). Place samplers in the river about 6-8 weeks prior to projected study. It is important to place samplers in inconspicuous locations, and somewhat protected from strong currents. Good locations include

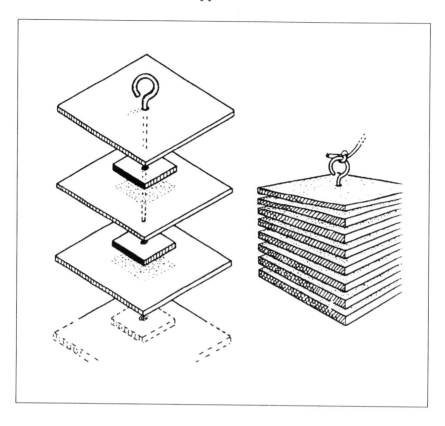

the downstream side of bridge abutments, along pilings, and other protected locations. This sampler is also discussed in Chapter 6, "Benthic Sampling
and Devices."

Materials

- 3/8 inch (approx. 1 cm) thick large piece of fiberboard, or any material that is somewhat water resistant and has a textured surface
- hacksaw
- ruler
- large eyebolt
- drill

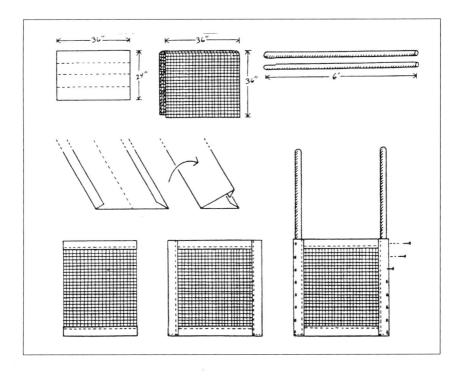

Directions

1. Cut squares of fiberboard
2. Drill a hole through the center of 10 squares
3. Cut smaller spacers.
4. Run the eyebolt through all of the spacers and the squares and tighten nut.
5. Attach heavy nylon to the eyebolt and attach to something fixed like a pier or bridge abutment.

▼
EQUIPMENT CONSTRUCTION PROJECT #3
Kick Net

Used in collecting benthic macroinvertebrates and fish. Only application is in fairly shallow rivers and streams. See activities 7.15 and 7.16 in Chapter 7. Also discussed in Chapter 6, "Benthic Sampling and Devices."

Materials

- 1 meter by 1 meter piece of nylon screening
- 4 strips of heavy canvas (15 cm x 1 meter)
- 2 broom handles or wooden dowels (2 meters long)
- finishing nails
- thread
- sewing machine
- hammer
- iron and ironing board

Directions

1. Fold nylon screen in half (1 meter by 1 meter)
2. Fold edges of canvas strips under 12 mm, and press with iron.
3. Sew 2 strips at top and bottom and then use other 2 strips to make casings for broom handles or dowels on left and right sides. Sew bottoms of casings shut.
4. Insert broom handles or dowels into casings and nail into place with finishing nails.

▼
EQUIPMENT CONSTRUCTION PROJECT #4
Surber Sampler

The Surber sampler is a semi-quantitative approach to sampling invertebrates. It is used most in waters less than 30 cm deep. Discussed in Chapter 6, "Benthic Sampling and Devices."

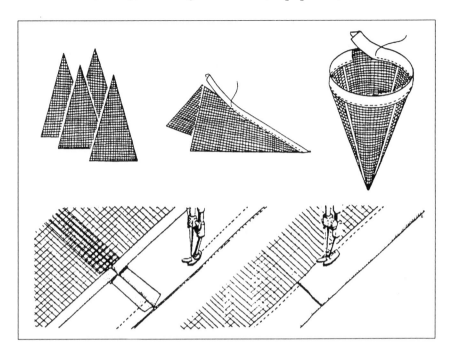

Materials

- 8 straight 30 cm wood or aluminum pieces (approx. 5 mm x 25 mm x 30 cm)
- 2 collapsible right angle corner braces (10-15 cm long)
- 36 nuts and bolts (6 mm in diameter and 12 mm long)
- 8 right angle braces (50 mm)
- 2 brass hinges (25-50 mm wide)
- 12 brass nuts and bolts for hinges
- 4 pieces of nylon netting (33 cm x 60 cm)
- 10 cm x 127 cm piece of heavy canvas (or several pieces sown together)
- bias tape
- hacksaw or wood saw
- screwdriver
- scissors
- drill with 5 mm and 6 mm bits
- sewing maching
- thread

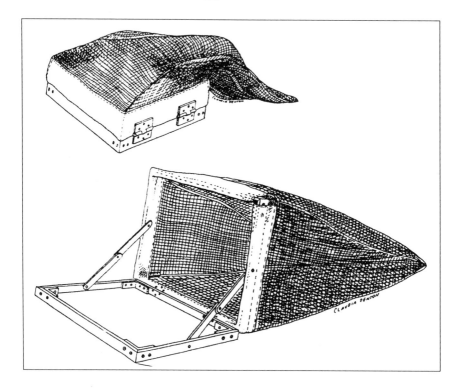

- adjustable wrench
- ruler
- pencil

Directions

1. Make two squares. Use four 30 cm pieces of wood or aluminum to make each square. Mark bolt positions using ruler and pencil. Drill holes for bolts to go through. Use right angle braces to put two frames together and use wrench and screwdriver to tighten down the nuts. Leave corner of one frame unbolted to slip the net on.

2. Cut netting into 4 triangular pieces (60 cm high with 33 cm bases.)

3. Sew edges of 4 pieces of netting together using bias tape as shown.

4. To make net casing, sew 10 cm ends of canvas together to form a wide cylinder. Fold in half and sew the edges of the

casing to netting, leaving an opening in casing to slip it onto the frame. Finished net should be 66 cm long.

5. Slip net on the unbolted corner of frame, put right angle brace in place and tighten down the nuts.

6. Lay two frames beside one another and position two hinges. Use a pencil to mark where you are going to drill. Drill holes and attach hinges. Make sure two frames fold flat.

7. Open frames to a right angle and position collapsible right angle braces. Mark where you are going to drill with pencil, drill holes, and attach.

▼

EQUIPMENT CONSTRUCTION PROJECT #5

D-Frame Net

Widely used in riffle, or shallow areas of streams and rivers. Used in activities 7.15 and 7.16 in Chapter 7. Discussed in Chapter 6, "Benthic Sampling and Devices."

Materials

- 4 pieces of nylon netting (25 cm x 30 cm)
- 25 mm bias tape or fabric scraps (1 meter long)
- thread
- scissors
- sewing machine
- wire coat hanger
- wire cutters
- drill with 6 mm wood bit
- broom handle or wooden dowel (120 cm long)
- pliers
- duct tape

Directions

1. Cut netting into four triangular pieces (25 cm high with 30 cm bases) as shown and sew together.

2. Cut a 120 cm strip of bias tape or fabric to make casing and sew onto net opening, leaving casing open to insert wire frame.

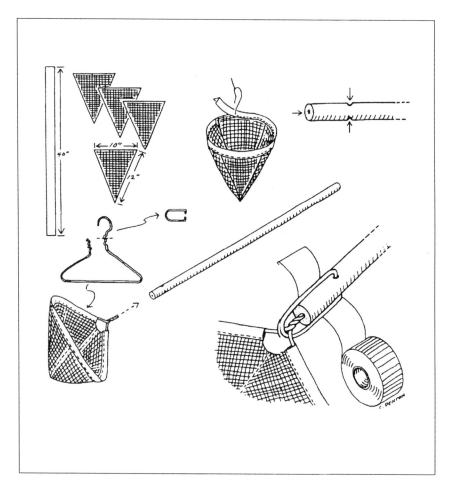

3. Take a wire coat hanger and untwist, slip into net casing, and retwist. Cut stem to 50 mm with wire cutters.

4. Drill hole in a broom handle or dowel and insert the stem as shown.

5. Take one of the remaining pieces of coat hanger and cut and bend it into a U-shape as shown.

6. Drill two shallow hole in handle, put U-shaped piece into position, push into holes as shown, and wrap with tape to secure handle.

▼

EQUIPMENT CONSTRUCTION PROJECT #6
Stream Plankton Sampler

Used in activity 7.13 to collect plankton samples.

Materials

- wire
- kitchen sieve (any size) with handle
- pantyhose (no holes)
- test tube or baby food jar

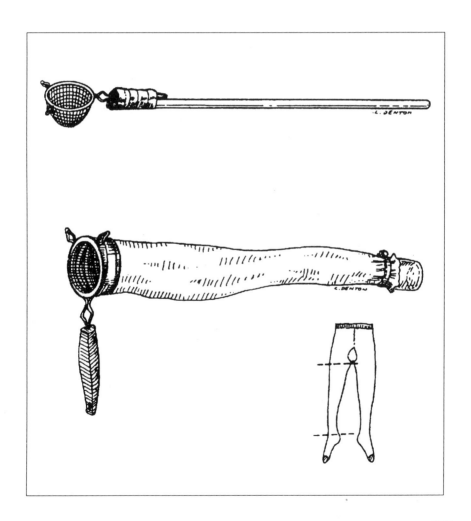

- heavy-duty rubber band
- electrical or duct tape
- scissors
- broom handle

Directions

1. Cut off one leg of pantyhose with scissors, then cut off foot end.
2. Attach larger end of pantyhose to kitchen sieve as shown. If sieve has metal loops, cut small slits in pantyhose and slip over the loops. Secure in place with tape.
3. Attach baby food jar to smaller end of pantyhose with rubber band.
4. Use tape to join the broom handle to sieve handle as shown.

▼
EQUIPMENT CONSTRUCTION PROJECT #7
Sampling Pans

Used in activities 7.15 and 7.16 in Chapter 7. The white background of these pans offers the greatest contrast to the normally dark benthic macroinvertebrates.

Materials

- opaque plastic bottles or milk jugs
- scissors or sharp knife

or:

- aluminum pie pans
- white enamel spray paint
- newspaper

Directions

1. Cut off top portion of opaque plastic bottles or milk jugs as shown; this leaves a 50 mm deep sampling dish.
2. Place aluminum pie pan on newspaper and paint with white enamel spray paint.

▼

EQUIPMENT CONSTRUCTION PROJECT #8
Underwater Viewer

This device may be helpful in determining amount of algal cover on the rocks, or in assessing the types of substrates present in the river. May be helpful during Activity 7.9 in Chapter 7, "Physical Characteristics."

Materials

- any size plastic, metal, or wooden bucket
- 6 mm thick plexiglass piece (size to fit bottom of bucket)
- sabre saw
- hand saw or tin snips
- silicone sealant or duct tape

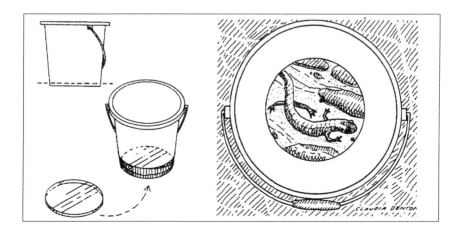

Directions

1. Use sabre saw to cut plexiglass into a circle to fit diameter of bucket.

2. Use regular saw or tin snips to cut off bottom of bucket

3. Use silicone sealant or duct tape to attach plexiglas to bottom of bucket.

▼

EQUIPMENT CONSTRUCTION PROJECT #9

Secchi Disk

The Secchi disk is used to measure turbidity in lakes, ponds, and slow-moving rivers. See photo of a Secchi disk in Fig. 1.4, Chapter 1.

Materials

- 20 cm. diameter lid from paint can or other metal can
- paint (black and white)
- eyebolt and nut
- rope
- drill

Directions

1. Cut out lid from a container; lid should be about 20 cm across.

2. Paint the lid in four quadrants, black, white, black white.

3. Drill a hole through the lid and insert the eyebolt.

4. Tie the rope onto the eyebolt.

▼
EQUIPMENT CONSTRUCTION PROJECT #10

Fecal Coliform Incubator

(Original design by: Chuck Dvorsky, Wes Halverson and students. Modified design by: Gordon Yamazaki.) Assembly price: $55–$60 (US)

Materials

1. *Cooler:* Any cooler which holds 20-28 liters of water. A flat top is preferred for ease of mounting components.

2. *Water pump:* Any pump which can be completely submersed for extended periods of time, generally an under gravel filter pump. The Powerhead 201 pumps approximately 480 L/hr on its highest setting. A larger pump would be needed for a larger cooler.

3. *Heater:* The Penn Plax heater is a 50 watt model and is recommended by the manufacturer for 50 gallon (~190 liter) aquariums. Therefore, it is powerful enough for any cooler application. Any heater with an adjustable thermostat will do but one with at least 35 watts is recommended to save heating time.

Design and Construction

1. *Water Pump:* Mount the water pump on the inside of the cooler near the bottom where it will be completely submersed. Complete submersion decreases wear and tear on the pump. Orientate the pump so that it causes water to circulate throughout the cooler. The Powerhead 201 comes with a suction cup mount for easy mounting.

2. *Heater:* Most submersible heaters have a water line marked on them that indicates from what point the heater must be submerged in water for optimal performance. The Penn Plax heater has a water line very near the thermostat control. Therefore this heater works best if mounted *through the side* of the cooler. The heater should be placed parallel with the floor of the cooler and about one inch (2-3 cm) above the floor.

To mount the heater through the side of the cooler, first measure and mark the position; then drill a hole though the side of the cooler. (A "doorknob" drill bit works best for this.) Always start drilling a smaller hole than you think you need

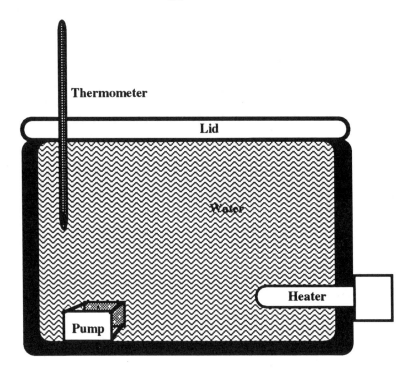

and slowly move up from there. Insert the heater through the hole. Once the heater is firmly in place, seal the edges of the hole with a flexible, waterproof bath caulk or a silicon sealer designed for aquarium use. Use sealant liberally and allow to dry thoroughly (36 hrs, unless otherwise specified).

3. *Thermometer:* Use a 1/4 inch bit to drill a hole through the top of the cooler. Mount the thermometer through the hole so that the bulb is about one inch below the surface of the water. Be sure to mount the thermometer away from the heater.

Start-Up

Fill the incubator with water to overflowing. To lessen the start-up time, use warm water (less than 44.5°C). Start the water pump. Wait *15 minutes* before starting the heater. Turn on heater. If the heater light is not on, increase the setting until the light comes on. Leave the thermostat in this position until the heater/light goes off. Check temperature and increase the thermostat on setting if necessary. Repeat process until a stable temperature of 44.5°C is reached.

Appendix B

Handouts, Data Sheets and Surveys

CATCHMENT ASSESSMENT COVER SHEET

1. Evaluator's Name(s): _____

2. Date:_____

3. Country: _____

4. State, Region or Province: _____

5. Nearest City, Town or Village: _____

6. Catchment/Watershed:_____

7. Name of Stream or River (or other water body): _____

8. Observation Points: _____

9. Survey Reach Length (m): _____

10. Reason for Survey and Goals of Catchment Assessment Activities:

_____

_____

_____

_____

_____

_____

_____

_____

_____

_____

_____

_____

_____

_____

ACTIVITY 1.4 DATA SHEET
MAPPING A CATCHMENT

Watercourse/Water Bodies

1. Area of Catchment: _____

2. Number of branches within the catchment: _____

3. Watercourse name(s): _____

4. Types of watercourses:

   ☐ ephemeral   ☐ intermittent   ☐ perennial

5. Average watercourse gradientv _____m/km

6. Discharge: _____m³/sec

Local Land Use in Catchment

7. Land uses in catchment (check all that apply in appropriate categories and add percentages if known or use best estimate of percentage):

   Agriculture:

   ☐ Pasture/grazing   ☐ Row Crops   ☐ Orchards

   ☐ Dry-Cropping   ☐ Agroforestry   ☐ Pesticide/Herbicide Use

   ☐ No Till   ☐ Till   ☐ Grains (wheat, oats, etc)

   ☐ Other (explain) _____

   Total Agricultural Area Percentage _____

   Urban/Suburban areas:

   ☐ Residential   ☐ Commercial   ☐ Golf Courses

   ☐ Other (explain) _____

   Total Urban/Suburban Area Percentage _____

Industrial areas (specify):

_____

_____

Total Industrial Area Percentage _____

Mining:

☐ Surface   ☐ Deep

☐ Other (explain)_____

Total Mining Area Percentage: _____

Logging:

☐ Clearcut   ☐ Selective cut

☐ Other (explain)_____

Total Logging Area Percentage: _____

Grassland:

☐ Grassland

Total Grassland Area Percentage: _____

Forested land:

☐ Forested land

Total Forested Land Area Percentage: _____

Other uses (explain; for example, sanitary landfill):

_____

_____

_____

_____

Total Area Not Elsewhere Classified: _____

(NOTE: Percentages must add up to 100)

ACTIVITY 2.2 DATA SHEET

BANK AND RIPARIAN VEGETATION EVALUATION

Describe below the natural physical surroundings of the catchment you are observing.

Bank vegetation:

☐ Barren   ☐ Grasses   ☐ Brush

☐ Deciduous   ☐ Conifer   ☐ Other

Species of bank vegetation:

Herbaceous	Woody

Use the space below for drawing:

BANK VEGETATION   SCORE _____

    4 (excellent)   Native vegetation in undisturbed state
    3 (good)   Mostly native vegetation mildly disturbed
    2 (fair)   Native vegetation moderately disburbed
    1 (poor)   Exotics, native vegetation severly disturbed

Verge vegetation:

☐ Barren   ☐ Grasses   ☐ Brush   ☐ Deciduous   ☐ Conifer

☐ Other_____

Species of verge vegetation:

Herbaceous   Woody

_____   _____

_____   _____

_____   _____

_____   _____

_____   _____

_____   _____

_____   _____

Use the space below for drawing:

VERGE VEGETATION   SCORE _____

4 (excellent)	Native vegetation present/canopy intact
3 (good)	Mostly native vegetation/canopy virtually intact
2 (fair)	Native vegetation clearly disturbed
1 (poor)	Exotics/cleared land or urban development

ACTIVITY 2.3 DATA SHEET

BANK EROSION/STABILITY EVALUATION

1. Estimate the percentage of bare soil: _____%

2. Bank slope:

   ☐ Steep   ☐ Moderate   ☐ Slight

3. Bank stability:

   ☐ Stable   ☐ Slightly eroded   ☐ Moderately eroded

   ☐ Severely eroded

EXTENT OF SLUMPING AND MOVEMENT   SCORE _____

   4 (excellent) No movement
   3 (good)   Slight movement on the banks
   2 (fair)   Moderate bank collapses
   1 (poor)   Severe bank failure with extensive cracking and fall-ins

AMOUNT OF BANK EROSION   SCORE _____

   4 (excellent) Stable, no sign of any bank erosion
   3 (good)   Very occasional and very local erosion
   2 (fair)   Some erosion evident
   1 (poor)   Extensive erosion

ACTIVITY 3.1 DATA SHEET

PHYSICAL CHARACTERISTICS

1. Location Point: _____

2. Stream Type:

   ☐ Straight   ☐ Meandering   ☐ Braided

   ☐ Channelized   ☐ Pool/Riffle

3. Today's Weather: _____

4. Last Precipitation:

   Date _____

   Amount (mm) _____

   Duration (hours) _____

5. Recent Weather: _____

   (For data requiring more than one sample, each sample should be taken by different students/participants and the samples should then be averaged for the final figure.)

   Average   Sample 1 Sample 2 Sample 3 Sample 4 Sample 5

6. Air Temp:

   _____   _____   _____   _____   _____   _____

7. Water Temp:

   _____   _____   _____   _____   _____   _____

Physical Characteristics

   Average   Sample 1 Sample 2 Sample 3 Sample 4 Sample 5

8. Stream width (m):

   _____   _____   _____   _____   _____   _____

9. Stream depth (m):

   _____   _____   _____   _____   _____   _____

10. Surface velocity (m/sec):

   _____   _____   _____   _____   _____   _____

11. Bankfull width (m):

	Average	Sample 1	Sample 2	Sample 3	Sample 4	Sample 5
(a)	_____	_____	_____	_____	_____	_____
(b)	_____	_____	_____	_____	_____	_____
(c)	_____	_____	_____	_____	_____	_____
(bw)	_____	_____	_____	_____	_____	_____

12. Channel slope (m/km):

_____ _____ _____ _____ _____ _____

13. Channel cross-section:

☐ rectangular ☐ U-shaped ☐ V-shaped ☐ other _____

14. Watercourse bottom, predominant type:

Inorganic:

☐ bedrock   ☐ boulder

☐ cobble   ☐ gravel

☐ sand   ☐ silt

☐ clay

☐ other _____

Organic:

☐ muck-mud (black, very fine organic)

☐ pulpy peat (unrecognizable plant parts)

☐ fibrous peat (partially decomposed plants)

☐ detritus (sticks, wood, coarse plant material)

☐ logs, limbs

Substrate composition:

% inorganic _____   % organic _____

15. Frequency of flooding (if known or best estimate):

☐ none   ☐ rare (10 to 20 years)

☐ occasional (5 to 10 years)   ☐ frequent (1 to 5 years)

☐ seasonal

16. Watercourse channel alteration including dates (if known):

☐ dredged _____   ☐ channelized _____

☐ dam/weir _____   ☐ wetland drainage ____

☐ other _____

ACTIVITY 3.2 DATA SHEET
PRIMARY USES AND IMPAIRMENTS

1. Population served: _____

2. Catchment primary uses by humans:

   ☐ Domestic drinking water supply

   ☐ Bathing

   ☐ Recreation:

       ☐ Swimming   ☐ Fishing

       ☐ Other (explain) _____

   ☐ Washing clothes

   ☐ Agricultural water supply:

       ☐ Irrigation   ☐ Livestock   ☐ Other (explain) _____

   ☐ Transportation:

       ☐ Motorized boats ☐ Non-motorized boats ☐ Commercial

   ☐ Industrial water supply

   ☐ Waste disposal

   ☐ Other uses (explain) _____

3. Water use impairments:

   ☐ No   ☐ Yes. The impairment(s) is/are due to (check all that apply):

   ☐ Agricultural runoff ☐ Livestock yards ☐ Cropland/ Pasture

   ☐ Inadequate or overloaded wastewater treatment facilities:

       ☐ primary   ☐ secondary   ☐ tertiary

   ☐ Logging runoff ☐ Mining runoff ☐ Industrial Discharge

   ☐ Golf courses ☐ Irrigation problems ☐ Housing

   ☐ Failing septic tanks ☐ Urban or other construction

   ☐ Other (explain) _____

ACTIVITY 3.3 DATA SHEET

WATER ODORS AND APPEARANCE

Odors

What smells does the watercourse have (check all that apply)?

	Water			*Soil*		
	Faint	Distinct	Strong	Faint	Distinct	Strong
Chemical:						
Chlorine	☐	☐	☐	☐	☐	☐
Sulfur (rotten eggs)	☐	☐	☐	☐	☐	☐
Musty:						
Decomposing straw	☐	☐	☐	☐	☐	☐
Moldy	☐	☐	☐	☐	☐	☐
Harsh:						
Fishy	☐	☐	☐	☐	☐	☐
Sewage	☐	☐	☐	☐	☐	☐
Earthy:						
Peaty	☐	☐	☐	☐	☐	☐
Grassy	☐	☐	☐	☐	☐	☐
Aromatic						
Spicy	☐	☐	☐	☐	☐	☐
Balsamic						
Flowery	☐	☐	☐	☐	☐	☐

☐ Other (explain) _____

☐ No unusual smells

Water Appearance

By visually observing the watercourse, what appearance does the water have (check all that apply)?

☐ Green ☐ Orange-red ☐ Foam ☐ Reds

☐ Blues ☐ Purples ☐ Blacks ☐ Milky/white

☐ Muddy/cloudy ☐ Multi-colored (oily sheen)

☐ Other (explain) _____

☐ No unusual colors

ACTIVITY 3.3 INFORMATION SHEET

WATER ODORS AND APPEARANCE (2)

Water Appearance Color Standards

Verbal descriptions of apparent color can be unreliable and subjective. If possible, use a system of color comparison that is reproducible. By using established color standards, people in different areas can compare these results. Match your sample to a color standard. Record the reference number of the color standard yielding the best match. Be sure to report the system of color standards used along with your observations.

Two color standard systems:

➤ Forel Ule Color Scale - good for offshore and coastal bay waters.

➤ Borger Color System - good for natural waters, in addition, colors of insects, larvae, algae, and bacteria.

To order: Ben Meadows Company
3589 Broad Street
Chamblee, GA 30341 USA
USA and Canada: 1(800)241-6401
Elsewhere: 1(404)455-0907
Fax: 1(404)457-1841

Forel-Ule Color Scale Order #224220   Price US $51.00

Borger Color System Order #224218   Price US $5.95

ACTIVITY 3.4 DATA SHEET

HABITAT ASSESSMENT

1. Notice the places around you where plants and animals could live. Check all those items below that apply to the area of the catchment that you are observing.

   ☐ Pool   ☐ Riffle/rapids   ☐ Wetlands

   ☐ Rocks   ☐ Log piles   ☐ Weed beds

   ☐ Undercut banks

   ☐ Human-made objects (pilings, bridges) specify _____

   ☐ Other (please describe) _____

2. Animals are an important part of a catchment ecosystem. In the spaces below, list the name (if known) of all the fish, reptiles, and birds that were seen during your observations today. If the name is unknown, try to draw a picture that best depicts the animals that were seen.

   BIRDS

   FISH

   REPTILES

General Comments: _____

_____

_____

_____

3. In the box below, draw a cross-sectional area of the water under investigation. Include vegetation growth on banks, shape of channel, etc.

HABITAT ASSESSMENT   **SCORE** _____

4 (excellent)	Bends present, 5–10 riffles in 10 meters, many snags
3 (good)	Bends present, 1–2 riffles in 10 meters, some snaps
2 (fair)	Occasional bend, 1–2 riffles in 50 meters, few snags
1 (poor)	Straight channel, riffles-pools absent, no snags

ACTIVITY 5.2 DATA SHEET

POLLUTION TOLERANCE INDEX (PTI)

Record the presence and estimate the number of each organism collected:

	Amount			
Group 1 (Index Value 4.0)	1–9	10–49	50–99	100 or more
Stonefly	☐	☐	☐	☐
Mayfly	☐	☐	☐	☐
Caddis Fly	☐	☐	☐	☐
Dobsonfly	☐	☐	☐	☐
Riffle Beetle	☐	☐	☐	☐
Water Penny	☐	☐	☐	☐
Gill Snail	☐	☐	☐	☐
Group 2 (Index Value 3.0)	1–9	10–49	50–99	100 or more
Damselfly	☐	☐	☐	☐
Dargonfly	☐	☐	☐	☐
Sowbug	☐	☐	☐	☐
Scud	☐	☐	☐	☐
Crane Fly	☐	☐	☐	☐
Clam	☐	☐	☐	☐
Group 3 (Index Value 2.0)	1–9	10–49	50–99	100 or more
Midge (excluding Blood Midges)	☐	☐	☐	☐
Black Fly	☐	☐	☐	☐
Flatworm	☐	☐	☐	☐
Leech	☐	☐	☐	☐
Water Mite	☐	☐	☐	☐
Group 4 (Index Value 1.0)	1–9	10–49	50–99	100 or more
Pouch Snail	☐	☐	☐	☐
Tubifex	☐	☐	☐	☐
Blood Midge	☐	☐	☐	☐
Maggot	☐	☐	☐	☐

CUMULATIVE INDEX VALUE　　　　　　　　　　**SCORE** _____

    4 (excellent)   23 or more
    3 (good)   17–22
    2 (fair)   11–16
    1 (poor)   10 or less

DISTINGUISHING CHARACTERISTICS OF COMMON MACROINVERTEBRATE TAXA GROUPS

INDEX VALUE 4.0

These organisms are generally pollution-intolerant. Their dominance generally signifies good water quality.

Stonefly Nymph (Order Plecoptera):

- 5–35 mm long, not including tail; sometimes up to 60 mm.
- Eyes: Widely separated.
- Antennae: Long and slender.
- Body: 1st three segments are hard on top; dorsally-ventrally compressed (flat).
- Abdomen: Ends in 2 tail filaments (cerci).
- Gills: Sometimes lacking; if present, at the base of the legs.
- Legs: Well developed; each ends in 2 claws.

Mayfly Nymph (Order Ephemeroptera):

- 3–20 mm long (not including tail).
- Antennae: Slender and long.
- Legs: 3 pairs, well developed, segmented, single claw or no claw on end.
- Cylindrical to flattened shape; streamlined.
- Abdomen: Series (usually 7 pairs) of gills arising from side.
- Gills may have gill covers with feather-like appearance or may be flat and spade-shaped.
- Usually 3 tail filaments; occasionally 2.
- Two-developing fore-wing pads are evident.
- Movement: Side to side.

Caddisfly Larva (Order Trichoptera):

- 2–40 mm in length.
- Eyes: Small and simple.
- Head: Hard-shelled head capsule.
- Body: Segmented; 1st 3 segments behind head may have hard-shelled plates on top surface.
- Legs: 3 pairs of thoracic legs are well-developed.
- Abdomen: Soft and cylindrical; end of abdomen has a pair of hooks at end of abdomen usually hooked and sharp.
- Gills either underneath or at the side of the body.
- Larva may live within portable cases (made of stone, plant, material) or nets.
- Movement: Series of loops.

Dobsonfly Larva (Order Megaloptera):

- 25–90 mm.
- Head capsule is hardened.
- Abdomen: 8 pairs of lateral filaments extending from abdominal segments; each filament has 2 segments.

- No tail, one pair of anal prolegs, each with 2 terminal hooks.
- 1st three body segments are hardened.

Water Penney Larva (Order Coleoptera, Family Psephenidea):

- 1 cm long.
- Flattened and disc-like, almost as broad as long.
- Dorsal plate-like expansions conceal head and legs from above, almost as broad as they are long.
- Highly adapted for adhering to stones.

Gilled Snail (Family Lymnaeidae):

- Spiral-shaped cone.
- Right-handed shell spirals. (To determine spiral direction, hold the shell in the vertical orientation with the aperture facing you. If the aperture opens on the right and the shell spirals clockwise, the shell is dextral, or right-handed. A sinistral shell spirals counter-clockwise.)
- This snail has gills and obtains oxygen from the surrounding water.

INDEX VALUE 3.0

These organisms can exist in a wide range of water quality conditions.

Damselfly Nymph (Order Odonata, Suborder Zygoptera):

- 20–120 mm long.
- Slender.
- 2 pair of wing pads.
- Cylindrical abdomen.
- 3 well-developed long leaf-like appendages protruding from the back end, instead of tail filaments.

Dragonfly Nymph (Order Odonata, Suborder Anisoptera):

- 20–120 mm long.
- Slender.
- Abdomen: broadens from the base, becoming wider toward the back end.
- No tail; abdomen ends in 3 short, wedge-shaped structures.
- Mouth: Hinged, shovel-like lower jaw that can be extended remarkably.

Aquatic Sowbugs (Order Isopoda):

- 5–20 mm long.
- Shape: Flattened.
- Legs: 7 pairs; first two pairs are modified for grasping.
- Abdomen: Segments are fused into a relatively short region.

Scud (Order Amphipoda):

- 5–20 mm long.
- Shrimp-like; body flattened.
- Well developed eyes.

- Thorax: 7 segments.
- Legs: 7 pairs; first pair are modified for grasping.
- Abdomen: 6 segments.
- Movement: Swims backward, on its side or back.

Crane fly (Order Diptera, Family Tipulidea):

- 15–100 mm long.
- Head: Retractable, only partially hardened.
- Oblong, cylindrical, somewhat tapered toward head.
- Abdomen: Last segment has a six-lobed plate or 2–6 short finger-like lobes.

Freshwater Clam (Class Pelecypoda, Family Pisisiidae):

- Bi-valve, two-piece shell.
- Shells usually oval, with concentric growth lines.

INDEX VALUE 2.0

These organisms are generally moderately tolerant of pollution. Their dominance usually signifies poor water quality.

Midge Larva (Order Diptera):

- 1 cm long.
- Legs: No jointed legs, as other true flies have.
- 2 pairs of pro-legs (fleshy and not jointed, short and stumpy), 1 pair just below head, 1 pair at back end.
- Body: Soft, slender, cylindrical; almost always curved in "C" or "S" shape.
- Color: Some common species are blood red (have hemoglobin which enables them to survive in water with little dissolved oxygen), others are not red.

Black fly Larva (Order Diptera, Family Simuliidae):

- Small size.
- Normally attached by its rear end to a substrate.
- Abdomen: Posterior part of the abdomen noticeably swollen.

Flatworm—Planaria (Order Tricladida, Family Planariidae):

- 1–30 mm long.
- Flattened bodies, elongate bodies
- Generally white, gray brown, or black
- Up to 30 mm
- They glide over submerged plants or stones
- Acutely triangular head

Leeches (Class Hirudinea):

- 5–400 mm long.
- Body: Many segmented; appears flattened dorsally-ventrally (top-to-bottom).
- Sucker on both ends of the body.
- May be patterned or brightly colored.
- Movement: By loops.

INDEX VALUE 1.0

These organisms are very tolerant of pollution. Their dominance usually signifies very poor water quality

Pouch or Lung Snails (Order Prosobranchia, Family Physidae):

- Tolerant of poor water conditions; they have lungs and come to the surface to breathe.
- Taller than they are wide.
- Pouch snails are sinistral or left-handed (see Physidae in Group 1 for explanation).
- Pouch snails do not have hemoglobin.

Sewage Worms (Family Tubificidae):

- 5 cm to 8 cm long
- Worm like and tapered, with sheath around base of body
- Body: soft, slender, cylindrical.

Blood Midge (Order Diptera):

- 1 cm. long.
- Legs: No jointed legs, as other true flies have.
- 2 pairs of pro-legs (fleshy and not jointed, short and stumpy), 1 pair just below head, 1 pair at back end.
- Body: Soft, slender, cylindrical; almost always curved in "C" or "S" shape.
- Color: Blood red (have hemoglobin which enables them to survive in water
- with little dissolved oxygen.

Rat-tailed Maggot (Family Syrphidae)

- Long posterior respiratory tube from once to several times the length of the body.
- Can live in polluted water, drawing air from the water surface through their respiratory tube.

Water Mite (Class Hydracarina)

- Small spider-like animals
- Eight legs
- Some have a red pigment.
- Swim freely in water

ACTIVITY 8.1 DATA SHEET

DISSOLVED OXYGEN

Date: _____

Testing Site Location: _____

Time: _____

Weather Conditions: _____

Names of team members:

_____   _____

_____   _____

_____   _____

_____   _____

Temperature Reading = _____ (in °C)

Number of drops _____ x 0.5 = _____ mg/liter → ____ % Saturation (from chart).

Number of drops _____ x 0.5 = _____ mg/liter → ____ % Saturation (from chart).

Number of drops _____ x 0.5 = _____ mg/liter → ____ % Saturation (from chart).

Number of drops _____ x 0.5 = _____ mg/liter → ____ % Saturation (from chart).

ADD UP ALL OF THE REASONABLE VALUES AND DIVIDE BY THE
NUMBER OF SAMPLES, (I.E. TAKE THE AVERAGE), TO GET THE
OFFICIAL MG/LITER (PPM) AND SATURATION LEVEL FOR YOUR SITE.

Dissolved Oxygen= _____mg/liter

% Saturation= _____

ACTIVITY 8.2 DATA SHEET
FECAL COLIFORM

Date: _____

Testing Site Location: _____

Time: _____

Weather Conditions: _____

Name of team members:

_____   _____

_____   _____

_____   _____

_____   _____

REMEMBER THAT ALL SAMPLE BOTTLES, PIPETTES, AND FILTRATION
SYSTEMS MUST BE STERILIZED BEFORE SAMPLING

1) Volume of sample _____(mL)

2)   → # _____ colonies after incubation

3)   conversion necessary → _____ Colonies/100 mL

1) Volume of sample _____(mL)

2)   → # _____ colonies after incubation

3)   conversion necessary → _____ Colonies/100 mL

1) Volume of sample _____(mL)

2)   → # _____ colonies after incubation

3)   conversion necessary → _____ Colonies/100 mL

IT IS IMPORTANT THAT YOU REPORT THE HIGHEST FECAL
COLIFORM VALUE RATHER THAN THE AVERAGE.

Official Reading =_____Colonies/100 mL

ACTIVITY 8.3 DATA SHEET

pH

Date: _____

Testing Site Location: _____

Time: _____

Weather Conditions: _____

Names of team members:

_____   _____

_____   _____

_____   _____

_____   _____

_____   _____

YOU MUST RUN THE TEST FOR pH IMMEDIATELY AFTER
SAMPLING BECAUSE CHANGES IN TEMPERATURE OF THE
SAMPLE CAN CHANGE THE MEASURED pH.

Values for test repetitions, (Try to take at least three measurements):

_____   _____   _____   _____

_____   _____   _____   _____

Take the most common value (mode) to report.

Official pH Reading = _____

ACTIVITY 8.4 DATA SHEET

BOD

Date: _____

Testing Site Location: _____

Time: _____

Weather Conditions: _____

Names of team members:

_____   _____

_____   _____

_____   _____

_____   _____

Temperature Reading = _____ (in °C)

First DO Test Results:

Number of drops _____ x 0.5 = _____ mg/liter → ____ % Saturation (from chart).

Number of drops _____ x 0.5 = _____ mg/liter → ____ % Saturation (from chart).

TAKE THE AVERAGE TO GET THE MG/LITER (PPM)
AND SATURATION LEVEL.

Dissolved Oxygen = _____ mg/liter   % Saturation = _____

DO Test Results AFTER FIVE DAYS

Number of drops _____ x 0.5 = _____ mg/liter → ____ % Saturation (from chart).

Number of drops _____ x 0.5 = _____ mg/liter → ____ % Saturation (from chart).

TAKE THE AVERAGE TO GET THE MG/LITER (PPM)
AND SATURATION LEVEL.

Dissolved Oxygen = _____ mg/liter   % Saturation = _____

BOD= _____ mg/liter (first reading)   *minus* _____ mg/liter (after 5 days) =

_____ mg/liter

ACTIVITY 8.5 DATA SHEET

TEMPERATURE

Date: _____

Testing Site Location: _____

Time: _____

Weather Conditions: _____

Names of team members:

_____   _____

_____   _____

_____   _____

_____   _____

REMEMBER: TEMPERATURE READINGS SHOULD BE IN °C.
TRY TO TAKE AT LEAST THREE SETS OF TEMPERATURE READINGS.

_____ Temp. downstream − _____ Temp. upstream = _____

_____ Temp. downstream − _____ Temp. upstream = _____

_____ Temp. downstream − _____ Temp. upstream = _____

_____ Temp. downstream − _____ Temp. upstream = _____

_____ Temp. downstream − _____ Temp. upstream = _____

_____ Temp. downstream − _____ Temp. upstream = _____

_____ Temp. downstream − _____ Temp. upstream = _____

TAKE THE MOST COMMON VALUE (MODE) TO REPORT.

Δ Temperature = _____

ACTIVITY 8.6 DATA SHEET

TOTAL PHOSPHATE

Date: _____

Testing Site Location: _____

Time: _____

Weather Conditions: _____

Names of team members:

_____   _____

_____   _____

_____   _____

_____   _____

_____ Reading in scale window÷ 50 = mg/liter Phosphate

_____ Reading in scale window÷ 50 = mg/liter Phosphate

_____ Reading in scale window÷ 50 = mg/liter Phosphate

TAKE AVERAGE IF YOU TAKE MORE THAN ONE READING.

Official Total Phosphate Reading = _____mg/liter

ACTIVITY 8.7 DATA SHEET

NITRATES

Date: _____

Testing Site Location: _____

Time: _____

Weather Conditions: _____

Names of team members:

_____   _____

_____   _____

_____   _____

_____   _____

REMEMBER TO DISPOSE OF THE TOXIC WASTE PRODUCTS
OF THIS TEST APPROPRIATELY.

_____ Reading in scale window x 4.4 = mg/liter Nitrate

_____ Reading in scale window x 4.4 = mg/liter Nitrate

_____ Reading in scale window x 4.4 = mg/liter Nitrate

TAKE THE AVERAGE IF YOU TAKE MORE THAN ONE READING.

Official Nitrate reading = _____mg/liter

ACTIVITY 8.8 DATA SHEET

TURBIDITY

Date: _____

Testing Site Location: _____

Time: _____

Weather Conditions: _____

Names of team members:

_____   _____

_____   _____

_____   _____

_____   _____

IF USING A SECCHI DISK (MEASUREMENTS IN FEET):

_____ (depth the Secchi disk disappears + depth disk reappears) ÷ 2 = _____feet

_____ (depth the Secchi disk disappears + depth disk reappears) ÷ 2 = _____feet

_____ (depth the Secchi disk disappears + depth disk reappears) ÷ 2 = _____feet

IF USING NTU'S, JTU'S OR FEET VALUES FOR A TEST REPETITION,
TRY TO TAKE AT LEAST THREE MEASUREMENTS.

TAKE THE MOST COMMON VALUE (MODE) TO REPORT.

Official Reading = _____   Feet Turbidity

ACTIVITY 8.9 DATA SHEET
TOTAL SOLIDS

Date: _____

Testing Site Location: _____

Time: _____

Weather Conditions: _____

Names of team members:

_____   _____

_____   _____

_____   _____

_____   _____

Weight of 300mL beaker WITH RESIDUE = _____ grams

− Weight of empty 300mL beaker BEFORE drying = _____ grams

Weight of residue = _____ grams

Formula for determining total solids is:

$$\frac{\text{Weight of residue (in grams)}}{\text{Volume of sample (in mL)}} \times \frac{1000 \text{ mg}}{1 \text{ gram}} \times \frac{1000 \text{ mL}}{1 \text{ liter}} = \text{_____ mg/liter}$$

Official Reading = _____ mg/liter

CHEMICAL REACTION EQUATIONS FOR THE CHEMICAL TESTS

Dissolved Oxygen Test

This tests the amount of O_2 actually in the water.

The first pillow you add, pillow #1, contains manganous sulfate powder. $MnSO_4$.

This is a compound containing Mn^{2+} ionically bonded to SO_4^{2-}. When added to water, the bond breaks, and $MnSO_4$ ionizes.

$$MnSO_4 \rightarrow \quad Mn^{2+} + SO_4^{2-}$$

Mn^{2+} ions add to the dissolved oxygen in the water, causing an oxidation-reduction reaction, as shown below:

$$2Mn^{2+} + O_2 \rightarrow 2Mn^{4+} + O^{2-} \text{ (brown-orange solid)}$$

This reaction works in a base, which donates hydroxide ions, forming a Mn^{4+} hydroxide, $MnO~OH)_2$. This product is a solid, which is the brownish-orange floc seen in the water.

Pillow #2, containing alkaline iodide azide, makes the solution basic (alkaline means basic). When pillow #3, containing sulfamic acid, is added, I⁻ from pillow #2 reacts with the iodide in solution, which starts out as I-, reacts with the precipitate in this acid to give I_3^- (I2 + I- in solution) and Mn^{2+} ions in water. The iodine, I_3^- , gives the solution a brown color.

$$Mn^{4+} + 3I^- \rightarrow Mn^{2+} + I_3^- \quad \text{(brown)}$$

The number of I_3^- ions formed is proportional to the amount of dissolved oxygen in the system, so if we determine how much I_3^- is present we can tell how much O_2 was in the water. Titrating the iodine with sodium thiosulfate causes a reaction which turns the solution from yellow to clear.

pH Test

If your water sample is acidic (pH<7), the following oxidation-reduction reaction will take place in he indicator (In):

$$H_2O + HIn \div \quad H_3O^+ + In^-$$

If your water sample is basic (pH>7), then this oxidation-reduction reaction will take place:

$$In + H_2O \quad \div \quad InH^+ + OH^-$$

In the first equation, the indicator acts as a weak acid, whereas in the second equation it acts as a weak base.

CHEMICAL REACTION EQUATIONS FOR THE CHEMICAL TESTS (2)

Nitrate Test

Adding the Nitra Ver 6 powder causes an oxidation-reduction reaction in which nitrate (NO_3-) is converted to nitrite (NO_2-), as shown below:

$$NO_3\text{-} + Cd + 2H^+ \rightarrow \qquad NO_2\text{-} = Cd^{2+} + 2H_2O$$

Cadmium loosens electrons, and so is said to be oxidized. Nitrogen gains these electrons, so is said to be reduced.

Adding a variety of acids to the nitrite creates salts of different colors. The color intensity of the final salt indicates how much nitrate was originally in the water.

Total Phosphate

Phosphate, PO_4^{3-} exists in water in either organic or inorganic stable compounds, not as unstable ions. To test for PO_4^{3-}, we must convert these stable compounds to the acid H_3PO_4. This involves an oxidation-reduction reaction, which takes place when you add 2 mL sulfuric acid to your water sample:

$$2\ PO_4^{3-} \text{ in water} + 3\ H_2SO_4 \rightarrow 3\ SO_4^{2-} + 2\ H_3PO_4$$

Adding NaOH causes an acid-base reaction, forming H_2PO_4-:

$$H_3PO_4 + NaOH \rightarrow \qquad H_2O + Na + H_2SO_4$$

The contents of the PhosVer 3 Reagent Powder Pillow react with $Na + H_2SO_4$ to create a compound of blue-violet color. The intensity of this color indicates the extent of PO_4^{3-} in the water.

WATER QUALITY MONITORING QUANTITATIVE ANALYSIS

VERGE VEGETATION   SCORE _____
- 4 (excellent)   Vegetation present and canopy intact
- 3 (good)   Vegetation and canopy virtually intact
- 2 (fair)   Vegetation clearly disturbed
- 1 (poor)   Cleared land or urban development

BANK VEGETATION   SCORE _____
- 4 (excellent)   Vegetation in undisturbed state
- 3 (good)   Vegetation mildly disturbed
- 2 (fair)   Vegetation moderately disturbed
- 1 (poor)   Vegetation severely disturbed

BANK BARE SOIL (PERCENT)   SCORE _____
- 4 (excellent)   0–10
- 3 (good)   11–40
- 2 (fair)   41–80
- 1 (poor)   81–100

BANK EROSION   SCORE _____
- 4 (excellent)   Stable, no sign of any bank erosion
- 3 (good)   Very occasional and local erosion
- 2 (fair)   Some erosion evident
- 1 (poor)   Severe bank failure with extensive cracking and fall-ins

BANK SLUMPING AND MOVEMENT   SCORE _____
- 4 (excellent)   No movement
- 3 (good)   Slight movement on the bank
- 2 (fair)   Moderate bank collapses
- 1 (poor)   Severe bank failure with extensive cracking and fall-ins

BENDS AND RIFFLES   SCORE _____
- 4 (excellent)   Bends present, 5–10 riffles in 10 meters, many snags
- 3 (good)   Bends present, 1–2 riffles in 10 meters, some snags
- 2 (fair)   Occasional bend, 1–2 riffles in 50 meters, few snags
- 1 (poor)   Straight channel, riffles/pools absent, no snags

PHYTOPLANKTON   SCORE _____
- 4 (excellent)   High diversity of phytoplankton (blue-green, green diatoms, flagellates)
- 3 (good)   Alert level I (500–2000 potentially toxic blue-green algal cells/mL)
- 2 (fair)   Alert level II (2000–15,000 potentially toxic blue-green algal cells/mL) concern for drinking water supplies
- 1 (poor)   Alert level III (greater than 15,000 potentially toxic blue-green algal cells/mL) can cause death and illness in cattle and humans

MACROPHYTE RIVER COVER   SCORE _____

4 (excellent)	Patches of surface and underwater plant cover (<10%), abundant overhanging vegetation
3 (good)	Some surface and underwater plant cover (10–30%), some overhanging vegetation
2 (fair)	Abundant surface and underwater plant cover (10–50%), little overhanging vegetation
1 (poor)	Choked surface and underwater plant cover (50–100%), no overhanging vegetation

SEQUENTIAL COMPARISON INDEX   SCORE _____
(Benthic Macroinvertebrates)

4 (excellent)	0.9–1.0
3 (good)	0.6–0.89
2 (fair)	0.3–0.59
1 (poor)	0.0–0.29

POLLUTION TOLERANCE INDEX   SCORE _____
(Benthic Macroinvertebrates)

4 (excellent)	23 and above
3 (good)	17–22
2 (fair)	11–16
1 (poor)	10 or less

EPHEMEROPTERA, PLECOPTERA, TRICOPTERA   SCORE _____
(Benthic Macroinvertebrates)

4 (excellent)	more than 15 families
3 (good)	12–15 families
2 (fair)	8–12 families
1 (poor)	less than 8 families

DISSOLVED OXYGEN LEVELS (% SATURATION)   SCORE _____

4 (excellent)	91–110
3 (good)	71–90
2 (fair)	51–70
1 (poor)	<50

FECAL COLIFORM (PER 100 ML)   SCORE _____

4 (excellent)	<50 colonies
3 (good)	51–200 colonies
2 (fair)	100–1,000 colonies
1 (poor)	>1,000 colonies

pH (UNITS)   SCORE _____
    4 (excellent)   6.5–7.5
    3 (good)   6.0–6.4, 7.6–8.0
    2 (fair)   5.5–5.9, 8.1–8.5
    1 (poor)   <5.5, >8.6

BIOCHEMICAL OXYGEN DEMAND (PPM)   SCORE _____
    4 (excellent)   <2
    3 (good)   2–4
    2 (fair)   4.1–10
    1 (poor)   >10

TEMPERATURE CHANGE (DEGREES CELSIUS)   SCORE _____
    4 (excellent)   0–2
    3 (good)   2.2–5
    2 (fair)   5.1–9.9
    1 (poor)   10>

TOTAL PHOSPHATE (MG/L)   SCORE _____
    4 (excellent)   0–1
    3 (good)   1.1–4
    2 (fair)   4.1–9.9
    1 (poor)   >10

NITRATES (MG/L)   SCORE _____
    4 (excellent)   0–1
    3 (good)   1.1–3 (good)
    2 (fair)   3.1–5
    1 (poor)   >5

TURBIDITY (NTUS/FEET OR CM)   SCORE _____

	NTUs	Secchi Disk
4 (excellent)	0–10	>3 feet
		>91.5 cm
3 (good)	10.1–40	1 foot to 3 feet
		30.5 cm to 91.5 cm
2 (fair)	40.1–150	2 inches to 1 foot
		5 cm to 30.5 cm
1 (poor)	>150	<2 inches
		<5 cm

TOTAL SOLIDS (MG/L)   SCORE _____
    4 (excellent)   <100
    3 (good)   100–250
    2 (fair)   250–400
    1 (poor)   >400

Appendix C
Glossary

Acid Rain: rainfall with a pH of less than 7.0. One source is the combination of rain and sulfur dioxide emissions which are byproducts of the combustion of fossil fuels.

Aerobic: Organisms able to live only in the presence of air or free oxygen.

Algae (pl.), alga: a collective term referring to several groups of simple photosynthetic plants, mostly microscopic, lacking roots, stems and leaves; they can be found in a variety of habitats; many species of algae exist as single cells, others form simple filaments or colonies and others exist as more complex structures like the larger seaweeds.

Algal bloom: Rapid growth of algae on the surface of lakes, streams, or ponds; stimulated by nutrient enrichment.

Alkali: Any strongly basic substance of hydroxide and carbonate, such as soda, potash, etc., that is soluble in water and increases the pH of a solution.

Alkaline: The presence of alkalies in water or soil in amounts sufficient to raise the pH value above 7.0.

Alluvial soil: A deposit of sand, mud, etc., formed by flowing water.

Anaerobic: Organisms able to live and grow only where there is no air or free oxygen, and conditions that exist only in the absence of air or free oxygen.

Animal waste: Either solid or liquid products, resulting from digestive or excretory processes, and eliminated from an animal's body.

Aquifer: Any geological formation containing water, especially one that supplies water for wells, springs, etc.

Glossary terms adopted with permission from: (1) *Water Words,* North Dakota State Water Conservation Service; (2) Bones, David, *Partners in Environmental Education Resources (PEER), Facilitator's Handbook* (Masters Thesis), University of Michigan, Ann Arbor, Michigan; (3) *Waterwatch Handbook,* Victoria, Australia; (4) Glossary of Perfetti, and Terrell, Charles, *Water Quality Indicator Guide: Surface Water,* 1991, Soil Conservation Service, U.S. Department of Agriculture.

Aquifer system: A series of (more or less) interrelated aquifers providing a source of ground water throughout a large area.

Arable: Suitable for cultivation.

Arid: Describe regions where precipitation is insufficient in quantity for most crops and where agriculture is impractical without irrigation.

Bank stabilization: Implementation of structural features along a stream-bank to prevent or reduce bank erosion.

Benthic region: The bottom of a body of water, supporting the benthos.

Benthos: All the plants and animals living on or closely associated with the bottom of a body of water.

Billabong: an old river meander that has been cut off and become isolated from the main channel.

Biochemical oxygen demand (BOD): A measure of the amount of oxygen removed from aquatic environments by aerobic microorganisms for the metabolic requirements. Measurement of BOD is used to determine the level of organic pollution of a stream or lake.

Biological community: All of the living things in a given environment.

Biological magnification: samples the process where the concentration of a material increases in the animals higher in the food chain due to the increasingly larger rates of consumption by the higher organisms.

Biome: An extensive community of plants and animals whose composition is determined by soil and climate.

Biota: The plant and animal life of a region.

Capillary movement: movement of water in saturated soil towards an area of drier soil; it can refer to movement in any direction, but upwards capillary movement is of most significance to the salinity problem.

Catchment: the area of land that is drained by a river and its tributaries; the watershed or dividing line between catchments is physically defined by mountains, crests of hills or the ridge of high ground.

Channelization: The artificial enlargement or realignment of a stream channel.

Chemical oxygen demand (COD): A measure of the amount of oxygen needed to oxidize organic and inorganic material present in water or sediment; a measure of the organic and inorganic pollutant level of sewage and industrial waste water.

Coliform bacteria: A group of organisms usually found in the colons of animals and humans. The presence of coliform bacteria in water is an indicator of possible pollution by fecal material.

Conductivity: a measure of the inorganic materials and inorganic ions in water; as total soluble salts form a major component, conductivity is used to measure salinity of water.

Confined aquifer: An aquifer bounded above and below by impermeable beds of rock or soil strata or by beds of distinctly lower permeability than that of the aquifer itself.

Confluence: The place where streams meet.

Cultural eutrophication: The process whereby human activities increase the amounts of nutrients entering surface waters, giving increased algal and other aquatic plant population growths, resulting in accelerated eutrophication of the watercourse or water body.

Degradation (river beds): The general lowering of the streambed by erosive processes, such as scouring by flowing water.

Delta: A nearly flat, often triangular, plain of deposited sand, mud, etc., between diverging branches of a river mouth.

Detritus: organic debris from decomposing plant and animals; in particular leaves, flowers and twigs are food for benthic macro-invertebrates.

Discharge: In the simplest form, discharge means outflow of water. The use of this term is not restricted as to course or location and it can be used to describe the flow of water from a pipe or from a drainage basin. Other words related to it are runoff, streamflow, and yield.

Dissolve: A condition where solid particles mix, molecule by molecule, with a liquid and appear to become part of the liquid.

Dissolved oxygen (DO): The amount of oxygen dissolved in water. Generally, proportionately higher amounts of oxygen can be dissolved in colder waters than in warmer waters. Dissolved oxygen is necessary for aquatic life and the oxidation of organic materials.

Drainage area: The land area contributing runoff to a stream or other body of water, and generally defined in terms of acres or square miles.

E. coli (Escherichia coli): one of the species of bacteria in the fecal coliform group; it is found in large numbers in the gastro-intestinal tract and feces of warm-blooded animals and humans; its presence in water is considered indicative of fresh fecal contamination.

Ecology: A community of animals, plants, and bacteria, and its interrelated physical and chemical environment.

Ecosystem: A community of animals, plants, and bacteria and its interrelated physical and chemical environment.

Effluent: The sewage or industrial liquid waste which is released into natural waters by sewage treatment plants, industry, or septic tanks.

Ephemeral: a stream that only flows after rain.

Escherichia coli (E. coli): A bacterium of the intestines of warm-blooded organisms, including humans, that is used as an indicator of water pollution for disease-producing organisms.

Estuary: an open drainage depression adjacent to the sea, typically at the mouth of a river, into which the tide ebbs and flows; tide movements accentuate erosion and continually modify the drainage channels within the estuary.

Eutrophication: A natural process whereby a watercourse or water body receives nutrients, resulting in increased growth of algae and other microscopic plants, possibly leading to a water body clogged with aquatic vegetation.

Evapo-transpiration: The loss of water from a land area through evaporation from the soil, and through plant transpiration.

Fauna: The entire animal population of a specific region and/or time.

Fertilizer: Any substance used to make soil or water more productive. Fertilizers may be commercially produced or be the result of animal or plant activities.

Flood plain: the area covered by water during a major flood; the area of alluvium deposits laid down during past floods

Flora: The entire plant population of a specified region and/or time.

Flow: The rate of water discharged from a source; expressed in volume with respect to time.

Ground Water: Subsurface water found in the zone of saturation.

Ground-water overdraft: The portion of ground-water withdrawals that exceeds recharge; sometimes called ground-water mining.

Ground-water recharge: The inflow to another aquifer.

Habitat: The native environment where a plant or animal naturally grows or lives.

Heavy metals: any element with an 'atomic number' larger than 20 that can be precipitated by hydrogen sulfide in acid solution; e.g., copper, cadmium, chromium, lead and mercury.

Herbicide: A type of pesticide, either a substance or biological agent, used to kill plants, especially weeds.

Hydrologic cycle: The constant circulation of water from the sea, through the atmosphere, to the land, and back to the sea by overland, subterranean, and atmospheric routes.

Hydrology: The science of waters of the earth; water's properties, circulation, principles, and distribution.

Impervious: Incapable of being penetrated by water.

Insecticide: A type of pesticide, either a substance or biological agent, used to kill insects or insect-like organisms.

Intermittent stream: A stream or reach of a stream that flows only at certain times of the year because losses from seepage or evaporation are greater than the available streamflow.

Lake: Any inland body of standing water, usually freshwater, larger than a pool or pond; a body of water filling a depression in the earth's surface.

Larvae (pl.), larva: the pre-adult form which differs distinctly from the sexually mature adult and usually requires an intermediate development stage (i.e., the pupa) before developing into the adult.

Lentic waters: standing water bodies such as lakes and ponds.

Leaching: The removal of soluble organic and inorganic substances from the topsoil downward by the action of percolating water.

Lotic: A moving body of water, like a river.

Macroinvertebrate: animals without a backbone and visible to the naked eye.

Macrophyte: literally 'big plant', used to describe water plants, either rooted or floating, other than microscopic algae.

Micro-organisms: either plant or animal, (e.g., algae or bacteria) that are invisible or barely visible to the naked human eye.

Mouth of stream: The point of discharge of a stream into another stream, a lake or the sea.

Natural flow: The flow of a stream as it would be if unaltered by upstream diversion, storage, import, export, or change in upstream consumptive use caused by development.

Nitrate (NO$_3$): a compound of nitrogen.

Nitrogen (N): a colorless, tasteless element usually occurring in the gaseous state. It forms approximately 80% of the earth's atmosphere and is essential for all organisms.

Non-point source pollution: Pollution discharged over a wide land area, not from one specific location.

Nutrients: Elements or compounds essential to life, including carbon, oxygen, nitrogen, phosphorus, and many others.

Nymph: young, sexually immature stage of certain insects, usually similar to the adult in form, which do not require an intermediate development stage between the nymph and adult.

Oligotrophic water body: A water body characterized by few nutrients entering the water body, few to no shoreline aquatic plants, and rarely any plankton blooms.

Organic matter: Plant and animal residues, or substances made by living organisms.

Parts per million (PPM): The number of "parts" by weight of a substance per million parts of water. This unit is commonly used to represent pollutant concentrations. Large concentrations are expressed in percentages.

Percolation: The movement of water downward through the subsurface to the zone of saturation.

Periphyton: plants and animals that are attached to submerged objects, such as rocks, macrophytes and tree debris; often microscopic is size.

Permeability: the ease with which water flows through a sediment or rock (also called hydraulic conductivity).

Pesticide: Any chemical or biological agent that kills plant or animal pests. Herbicides, insecticides, nematocides, miticides, algaecides, etc., are all pesticides.

pH: An expression of both acidity and alkalinity on a scale of 0:14, with 7 representing neutrality; numbers less than 7 indicate increasing acidity and numbers greater than 7 indicate increasing alkalinity.

Photosynthesis: The process by which plants manufacture their own food (simple carbohydrates) from carbon dioxide (CO$_2$) and water. The plant's chlorophyll-containing cells use light as an energy source and release oxygen as a byproduct.

Phytoplankton: Usually microscopic aquatic plants (sometimes consisting of only one cell).

Plankton: Small-to-microscopic, floating or freely swimming, aquatic plants and animals.

Pollution: (of water) when the level of concentration is high enough to impair water quality to a degree that has an adverse effect upon any beneficial use of the water.

Point source pollution: Pollutants discharged from any identifiable point, including pipes, ditches, channels, sewers, tunnels, and containers of various types.

Pond: A body of fresh or salt water, smaller than a lake, and where the shallow:water zone (light penetration to its bottom) is relatively large compared to the open water and deep bottom (no light penetration to the bottom).

Porosity: The ratio (usually expressed as a percentage) of the volume of openings in a rock to the total volume of the rock.

Potable: Water fit for human consumption.

Precipitation: Water falling, in a liquid or solid state, from the atmosphere to a land or water surface.

Primary waste treatment: The removal of suspended and floatable solids which will settle out of sewage and industrial wastes. Primary treatment plants generally remove 25 to 35 percent of the biological oxygen demand and 45 to 65 percent of the total suspended matter.

Pupae (pl.), pupa: a developmental stage in an insect's life cycle between the larvae and the adult

Reach: Any arbitrarily defined length of a stream.

Recharge zone: an area of land where the groundwater moves downward and water infiltrates from the surface into the geological formations below.

Riparian areas: Land areas directly influenced by a body of water. Usually have visible vegetation or physical characteristics showing this water influence. Stream sides, lake borders, and marshes are typical riparian areas.

Riffle: a section of river or stream with rapid, turbulent flow; generally shallow.

River: A natural stream of water of substantial volume.

River basin: A term used to designate the area drained by a river and its tributaries.

Runoff: The amount of precipitation appearing in surface streams, rivers, and lakes; defines as the depth to which a drainage area would be covered if all of the runoff for a given period of time were uniformly distributed over it.

Secondary waste treatment: Treatment (following primary treatment) generally removes 80 to 95 percent of the BOD and suspended matter.

Sediment: Fragmented organic or inorganic material derived from the weathering of soil, alluvial, and rock materials; removed by erosion and transported by water, wind, ice, and gravity.

Sedimentation: The deposition of sediment from a state of suspension in water or air.

Sheet erosion: The removal of fairly uniform layers of surface material from gently sloping land by rainfall and runoff water acting in continuous sheets of water.

Silt: Sedimentary particles smaller than sand particles, but larger than clay particles.

Stream: Any body of running water moving under gravity flow through clearly defined natural channels to progressively lower levels.

Streambank erosion: The wearing away of streambanks by flowing water.

Suspended solids (SS): Defined in waste management, these are small particles of solid pollutants that resist separation by conventional methods. SS (along with BOD) is a measurement of water quality and an indicator of treatment plant efficiency.

Thermal pollution: The impairment of water quality through temperature increase; usually occurs as a result of industrial discharges of coolant water.

Topographic maps: Maps with lines showing equal elevation of a region's relief; also showing natural features, including hills, valleys, rivers, and lakes; and built structures such as canals, bridges, roads, cities, etc.

Total dissolved solid (TDS): The quantity of dissolved materials in the water.

Total suspended solids: Solids, found in waste water or in a stream, which can be removed by filtration. The origin of suspended matter may be human and industrial wastes or natural sources such as silt.

Toxin: Any of a variety of unstable, poisonous compounds produced by some micro:organisms and causing certain diseases.

Tributary: A stream that contributes its water to another stream or body of water.

Turbidity: Cloudiness caused by the presence of suspended solids in water; an indicator of water quality.

Waste water: Water that carries wastes from homes, businesses, and industries; a mixture of water and dissolved or suspended solids.

Waste water treatment: Any of the mechanical or chemical processes used to modify the quality of waste water in order to make it more compatible or acceptable to humans and their environment.

Watershed: Area of land that contributes surface runoff to a given point in a drainage system.

Water table: The top of the zone of saturation.

Wetlands: Wetlands are areas where water is a controlling factor in the development of plant and animal communities. It may be standing water above the ground, or an underground water table that is close to the surface. Water may be present throughout the year or only during part of the year. Wetlands are often transitional areas between upland habitats and aquatic habitats. Other common names for wetlands are bogs, swamps, and marshes.

Zone of aeration: A subsurface zone of water in the earth's crust above the permanent groundwater level. The bottom of the zone of aeration is the water table; the top is the land surface.

Zone of saturation: A subsurface zone in which all the pores of the material are filled with ground water under pressure greater than atmospheric pressure.

Appendix D
Bibliography

Allan, J.D. 1995. *Stream Ecology: Structure and Function of Running Waters.* Chapman & Hall, London, UK.

American Public Health Association. 1990. *Standard Methods for the Examination of Water and Waste Water.* 16th ed. American Public Health Association, Inc., NY.

Anderson, N.H. and Wallace J.B. 1984. Habitat, Life History, and Behavioral Adaptations of Aquatic Insects. In: R.W. Merritt and K.W. Cummins (eds.). *An Introduction to the Aquatic Insects of North America* (2nd Ed.) Kendall/Hunt Publishing Co., Dubuque, IA.

Aspen Global Change Institute. 1992. *Ground Truth Studies Teacher Handbook,* Aspen, CO.

Australian Nature Conservancy Agency. 1994. *Waterwatch.* Department of Environment and Land Planning. Canberra, ACT, Australia.

Baltic Sea Program. 1994. *The Baltic Sea Project.* National Agency for Education, Stockholm, Sweden.

Bardwell, Lisa , Monroe, Martha and Tudor, Margaret. 1994, *Environmental Problem Solving: Theory, Practice and Possibilities in Environmental Education.* North American Association for Environmental Education, Troy, OH.

Baumgartner, A. and Reichel, E. 1975. *The World Water Balance: Mean Annual Global, Continental and Maritime Precipitation and Run-off.* Elseview Scientific Publishers, Amsterdam.

Beckinsale, R.P. 1969. River Regimes. In: *Water, Earth and Man.* R.J. Chorley (Ed.). Methuen, London, UK.

Beschta, R.L. and Platts, T. 1990 W.S. Morphological Features of Small Streams: Significance and Function. *Water Resources Bulletin* 22(3):369-379.

Bull, James, et. al. 1988. *Education in Action: A Community Problem-Solving Program for Schools.* Thomson-Shore, Dexter, MI.

Burch, J.B. 1985. *Handbook on Schistosomiasis and other Snail-Mediated Diseases in Jordan.* University of Michigan, Ann Arbor, Michigan, University of Lowell, Lowell, Massachusetts, The Ministry of Health, Amman, Jordan, and The University of Jordan, Amman, Jordan, Hashemite Kingdom of Jordan.

Burt, T.P. 1992. The Hydrology of Headwater Catchments. pp. 3-28. In: *The Rivers Handbook: Hydrological and Ecological Principles,* Vol. 1. P. Calow and G.E. Petts (Eds.). Blackwell Scientific Publications, Oxford, UK.

Caduto, M.J. 1990. *Pond and Brook: A Guide to Nature Study in Freshwater Environments.* 2nd edition., Prentice-Hall, NJ.

Cairns, John. 1992. *Restoration of Aquatic Ecosystems.* National Academy Press, Washington, D.C.

Cairns, J., Jr. and Dickson, K.L. 1971. A Simple Method for the Biological Assessment of the Effects of Waste Discharges on Aquatic Bottom-Dwelling Organisms. *J. Water Pollution Control Fed.,* 43:755-772.

Carling, P.A. 1992. Instream Hydraulics and Sediment Transport. In: *The Rivers Handbook: Hydrological and Ecological Principles,* Vol. 1. P. Calow and G.E. Petts (Eds.). Blackwell Scientific Publications. Oxford, UK.

Chutter, F. M. 1972. *An Empirial Biotic Index of the Quality of Water in South African Streams and Rivers.* Water Research Pergamon Press, Vol 6:19-30. Great Britain, UK.

Croft, P.S. 1986. *A Key to the Major Groups of British Freshwater Invertebrates.* Field Studies 6: 531-579, Field Studies Council, Shrewsbury College of Arts and Technology, Shrewsbury, UK.

Cummins, K.W. 1992. Invertebrates. In: *The Rivers Handbook: Hydrological and Ecological Principles,* P. Calow and G.E. Petts (Eds.), pp. 234-250. Blackwell Scientific Publications, Oxford, UK.

Cummins, K.W. et al. 1989. Shredders and riparian vegetation. *BioScience.* 39: 24-30.

de Lange, Esther. 1994. *Manual for Simple Water Quality Analysis.*

International Water Tribunal, Damrak, Amsterdam, The Netherlands.

Delong, M.D. and Brusven, M.A. 1991. Classification and Spatial Mapping of Riparian Habitat with Applications Toward Management of Streams Impacted by Nonpoint Source Pollution. *Environmental Management.* 15(4):565-571.

Dunne, T. and Leopold, L.B. 1978. *Water in Environmental Planning.* W.H. Freeman, San Francisco, CA.

Fiske, Steve, and Byrne, Jack 1987. *Key to the Freshwater Macroinvertebrate Fauna of New England.* River Watch Network, Montpelier, VT.

Global Rivers Environmental Education Network. 1992. *Investigating Streams and Rivers.* GREEN, Ann Arbor, MI.

Global Rivers Environmental Education Network. 1994. *Walpole Island First National Water Quality Monitoring and Environmental Education Handbook.* GREEN, Ann Arbor, MI.

Global Rivers Environmental Education Network. 1993. *Project del Rio Lesson Plans for a 16 -Day Water Quality Monitoring Project.* GREEN, Ann Arbor, MI.

Global Rivers Environmental Education Network. 1992. *GREEN Cross-Cultural Partners Activities Manual.* GREEN, Ann Arbor, MI.

Gordon, N.D., McMahon, T.A. and Finlayson, B.L. 1992 *Stream Hydrology: An Introduction for Ecologists.* John Wiley & Sons, Inc., Chichester, UK.

Government of Swaziland. 1992. *The Kingdom of Swaziland.* Jubilee Printing and Publishing, Mbabana, Swaziland.

Haralambous, Sappho. 1990. *The State of World Poverty: A Profile of Latin America and the Caribbean.* International Fund for Agricultural Development. New York University Press and Intermediate Technology Publications, N.Y.

Hartmann, D.L. 1994. *Global Physical Climatology.* Academic Press, San Diego, CA.

Hewlett, J.D., and Hibbert, A.R. 1967. Factors Affecting the Response of Small Watersheds to Precipitation in Humid Areas, pp. 275-290. In: *Forest Hydrology.* W.E. Sopper and H.W. Lull (Eds.).

Hilsenhoff, W.L. 1988. Rapid Field Assessment of Organic Pollution with a family level biotic index. *J. North American Benthological Society* 7(1): 65-68.

Hynes, H. B. N. 1976. The Biology of Plecoptera. *Annual Review of Entomology.* 21:135-153.

Hynes, H.B.N. 1970. *The Ecology of Running Waters.* University of Toronto Press, Toronto, Canada.

Karr, J.R. 1981. Assessment of Biotic Integrity Using Fish Communities. *Fisheries.* 6: 21-27.

Ketchum, Bostwick H. 1988. *The Waters Edge: Critical Problems of the Coastal Zone.* MIT Press, London, England, UK.

Keuneman, Herbert. 1989. *Sri Lanka.* APA Productions, Colombo, Sri Lanka.

Krueger, Henry O, Ward John P, and Anderson Stanley H. 1988. *A Resource Manager's Guide for Using Aquatic Organisms to Access Water Quality for Evaluation of Contaminants.* Fish and Wildlife Service, U. S. Department of the Interior, Research and Development, Washington, D.C.

Lenat, D.R. 1988. Water Quality Assessment of Streams Using a Qualitative Collection Method for Benthic Macroinvertebrates. *Journal of North American Benthological Society,* 7:222-233.

Lennox, Colin and Sue Lennox, 1994. *Water is Life—Student Environmental Congress Manual.* Global Rivers Environmental Education Network. Harbord, NSW, Australia.

Leopold, L.B. 1944. *View of a River.* Harvard University Press, Cambridge, MA.

McCafferty, P.W. 1981. *Aquatic Entomology: The Fisherman's and Ecologist's Guide to Insects and their Relatives.* Jones and Bartlett Publishers, Inc., CA.

Madison, Stafford, and Melissa Paly. 1994. *A World in Our Backyard: A Wetlands Education and Stewardship Program.* New England Interstate Water Pollution Control Commission. Environmental Media Center, Chappel Hill, NC.

Mitchell, Mark K. and William B. Stapp 1994. *Field Manual for Water Quality Monitoring.* Thomson-Shore Publishers, Dexter, MI.

Moffatt, Robert. 1992. *Marine Studies: A Course for Senior Students.* Wet Paper Publications, National Library of Australia, Australia.

Morisawa, M. 1985. *Rivers: Form and Process.* Longman, UK.

Needham, James G. and Paul Needham. 1982. *A Guide to the Study of Freshwater Biology.* Holden-Day, San Francisco, CA.

O'Donoghue, Robert, et. al. 1994. *SWAP: Guide for Practical Water Quality Monitoring.* Juta and Company, Printed and Bound by The Rustica Press, Ndabeni, Cape, South Africa.

Palmer, Mervin C. 1959. *Algae in Water Supplies.* Public Health Service, U.S. Department of Health, Education, and Welfare, Washington, D.C.

Pennak, R. W. 1978. *Fresh-Water Invetebrates.* John Wiley and Sons, New York.

Pickford, John. 1987. *Developing World Waters.* Grosvenor Press International, Loughborough, UK.

Plafkin, J.C., Porter, K.D. Gross, S.K. and Hughes R.M. 1989. *Rapid Bioassessment Protocols for Use in Streams and Rivers: Benthic Macroinvertebrates and Fish.* U.S. Environmental Protection Agency, EPA 440 4-89 001, May 1989.

Project Wild. 1987. *Aquatic Education Activity Guide.* Boulder, CO.

River Watch Network. 1990. *Guide to Macroinvertebrate Sampling for White River Headwaters Citizens Monitoring Group.* Vermont Department of Water Resources, Montpelier, VT.

Rohwedder, W.J. 1990. *Computer-Aided Environmental Education.*

North American Association for Environmental Education. Troy, OH.

Schlosser, I.J., and Karr, J.R. 1981. Riparian vegetation and channel morphology impact on spatial patterns of water quality in agricultural watersheds. *Environmental Management* 5(3):233-243.

Schuett-Hames, Pleus, Allan, Bullchild, Lyman and Hall, Scott. 1993. *Ambient Monitoring Program Manual.* Northwest Indian Fisheries Commission. Olympia, WA.

Schumm, S.A. and H.R. Kahn. 1972. Experimental study of channel patters. *Bulletin of the Geological Society of America* 83:1755-70.

South Australia Waterwatch. 1994. *Catchment Care and Water Quality Monitoring Manual for South Australia.* Department of Environment and Natural Resources, Adelaide, Australia.

Stapp, William. 1969. *The Concept of Environmental Education,* "The Journal of Environmental Education," 1:1, Fall.

Stapp, W.B. and A.E.J. Wals. 1944. *An Action Research Approach to Environmental Problem Solving.* In: Bardwell, L.V. Monroe, M.C. and M.T. Tudor, Environmental Problem Solving: 66. Troy, Ohio, NAAEE.

Stapp, William, Margaret T. Pennock and Timothy P. Donahue. 1996. *Cross Cultural Watershed Partners.* Kendall/Hunt Publishing Company, Dubuque, IA.

Stapp, William, Arjen Wals, and Sheri Stankorb. 1996. *Environmental Education for Empowerment.* Kendall/Hunt Publishing Company, Dubuque, IA.

Stapp, William, David Schmidt, and Andy Alm. 1996. *Investigating Streams and Rivers.* Kendall/Hunt Publishing Company, Dubuque, IA.

Swank, W.T., L.W. Swift, Jr. and J. E. Douglas. 1988. *Hydraulic Stream Ecology Associated with Forest Cutting, Species Conversions, and Natural*

Disturbances. pp. 297-312. In: Forest Hydrology and Ecology at Coweeta. W. T. Swank and D. A. Crossley, Jr. (Eds.). Springer-Verlag, New York.

Taiwan Environmental Protection Agency. 1992. *An Introduction to Environmental Protection in the Taiwan Area.* Government of the Republic of China, Taipei, Taiwan.

Terrell, Charles, and Perfetti, Patricia. 1989. *Water Quality Indicators Guide: Surface Waters* . Soil Conservation Service, The United States Department of Agriculture. Washington, D.C.

The World Conservation Union. 1991. *Caring for the Earth: A Strategy for Sustainable Living.* IUCN, Gland, Switzerland.

United Nations Educational, Scientific, and Cultural Organization. 1994. *Argentina: Taller del Agua Parana River.* Life Link Foundation, Rosario, Argentina.

U.S. Environmental Protection Agency. 1990. *Monitoring Lake and Reservoir Restoration.* Office of Water, Washington, D.C.

U.S. Environmental Protection Agency. 1991. *Manual for Citizens Volunteers—River Monitoring.* Washington, D.C.

U.S. Environmental Protection Agency. 1994. *Streamwatch Manual.*

Water Division, Seattle, Washington, D.C.

Ward, J.V. 1992. *Aquatic Insect Ecology: Biology and Habitat.* John Wiley & Sons, Inc., N.Y.

Washington, H.G. 1982. *Diversity, Biotic and Similarity Indices: A Review with Special Relevance to Aquatic Ecosystems.* CSIRO Energy and Earth Resources. Ryde, N.S.W., Australia.

Western Regional Environmental Education Council. 1993. *Wetlands.*

Project Wet, Boulder, CO.

Williams, D.D. and B.W. Feltmate. 1992. *Aquatic Insects.* CAB International, Wallingford, UK.

Williams, Robert and Cindy Bidlack. 1993. *Rivers Curriculum: Language Arts.* Rivers Curriculum Project, Southern Illinois University, Carbondale, IL.

World Resources Institute. 1991. *World Resources 1990-91.* A Report by the World Resources Institute in Collaboration with the United Nations Environment Programme and the United Nations Development Programme. Oxford University Press, UK.

Index

GREEN
Global Rivers Environmental Education Network
Global citizens sharing their concerns for water quality

GREEN is an innovative, action-oriented approach to education, based on an interdisciplinary watershed education model. GREEN's mission is to improve education through a global network that promotes watershed sustainability. It is a resource to schools and communities that wish to study their watershed and work to improve their quality of life.

GREEN works closely with educational and environmental organizations, community groups, businesses, and government across the United States and Canada, and in over 130 countries around the globe to support local efforts in watershed education and sustainability.

Students at North Farmington High School near Detroit detected elevated levels of bacterial contamination down river from a pipe exiting a City sewage pumping station. They presented their findings to the City Engineer, who acted quickly to repair the malfunctioning pump.

GREEN Watershed Education Model

The model involves the synthesis of content and process. Activities revolve around two key areas: watershed and water quality monitoring, and understanding changes and trends within the whole watershed.

GREEN participants collect and analyze real-life environmental data; study current and historical patterns of land and water usage within their watershed; share their data, concerns and strategies for action with others in the watershed and beyond; and develop concrete action plans to improve local water quality.

Key to the GREEN process is an emphasis on creating a learning community of teachers, students, parents, community groups, government, nongovernmental organizations, and businesses—whose members share a vision for watershed sustainability and possess the skills, knowledge, and motivation necessary to create change.

How GREEN Can Help You

GREEN publishes a quarterly, **international newsletter** and a **catalog of educational resources,** hosts a **World Wide Web** home page, sponsors watershed-wide and international **computer conferences** on EcoNet, and connects classrooms internationally in **cross cultural partnerships,** develops and disseminates **watershed education materials,** and provides **training and support** to local watershed education efforts.

GREEN is pioneering a **partnership approach** to educational development and environmental sustainability. Our partners include National Science Foundation, U.S. Department of Education, U.S. EPA, U.S. Fish and Wildlife Service, White House Office of Science and Technology Policy, U.N. Environment Programme, General Motors Corp., Key Bank, Bullitt Foundation, Institute for Global Communications, I*EARN, River Watch Network, TERC, Trout Unlimited, National Project WET, University of Michigan, Western Washington University, American Rivers, Inc, Atlantic Center for the Environment, and the Center for Watershed Protection. Joint research and development initiatives with our partners allow GREEN participants to benefit from the broad array of expertise represented within our network.

One GREEN student described to her father the adverse impacts of allowing his livestock direct access to the stream. He agreed to buy fencing material if she would help him build a fence. They proceeded to build the fence adjacent to the stream and followed up by planting trees and shrubs to establish a riparian corridor.

GREEN Workshops and Institutes

GREEN can help you start or enhance your watershed education program. We'll create and deliver a **Custom Workshop** with support materials to fit the goals of your program and participants. Topics include:

➤ Starting a Watershed Education Program
➤ Interdisciplinary Approaches to Watershed Education
➤ Water Quality Monitoring & Data Interpretation

- ➤ Land Use Analysis With Maps & Satellite Imagery
- ➤ Telecommunications in Watershed Education
- ➤ Student Problem Solving & Action Taking

All workshops provide hands-on experience in the GREEN approach for watershed education; a variety of support materials; a basic understanding of the complex relationships between land, water, and people; and activities that demonstrate the importance of protecting the environment and how individuals and communities can effect environmental change.

GREEN also hosts a series of regularly scheduled, introductory **Watershed Education Institutes** at locations across the U.S. and Canada. Please contact GREEN for details.

A local business man working with Project del Rio, an international GREEN program that links schools along the U.S.-Mexico border, is convinced that the GREEN model is "exactly the kind of program that enables a community to participate in the preservation and restoration of its environment."

On the Cutting Edge

GREEN is engaged in numerous initiatives to develop and refine curricular materials and to pilot **innovative approaches to education** such as school-community linkages, inquiry-based learning, action research, use of telecommunications and technology, and cross cultural learning. With our partners we are developing software for modeling environmental data, digital technology for field investigations, Internet-accessible environmental database technology, and low-cost methods for environmental monitoring.

A fundamental part of our research and development strategy is the suite of GREEN publications. GREEN publishes manuals, curriculum guides, and videos to support global watershed education. Our titles cover topics such as: water quality monitoring, action taking, and cross-cultural partnerships. Contact GREEN if you would like to receive our catalog of educational materials and water monitoring equipment.

GREEN Publications

- • **Field Manual for Water Quality Monitoring** details nine chemical/physical water quality tests and methods for biological monitoring. It also includes chapters on heavy metals testing, land use practices, action taking, and computer networking. The Field Manual is the foundation for GREEN's Educational Model.

The following books provide more in-depth information on components of their model.

- *Field Manual for Global Low Cost Water Quality Monitoring* provides a global perspective for watershed education. It includes activities to help readers understand key concepts and build skills. It provides handouts and instructions for making inexpensive equipment.
- *Investigating Streams and Rivers* is an activity guide that promotes an interdisciplinary approach to watershed education. It focuses on action taking and enhancing student involvement through computer networks.
- *Sourcebook for Watershed Education* provides detailed guidelines for the development of watershed-wide education programs, focusing on program goals, funding, and school-community partnerships. It contains a rich set of interdisciplinary classroom activities and outlines GREEN's educational philosophy.
- *Cross Cultural Watershed Partners: Activities Manual* contains activities for use in an intercultural watershed education program, and suggestions on how to structure a cross cultural exchange around watershed themes.
- *Air Pollution: Ozone Study and Action* moves students from awareness of air pollution and ozone to a point where they will be knowledgeable and empowered to make action to address problems in their own lives and communities.
- *Environmental Education for Empowerment* enables students, teachers, administrators, and others to effectively participate in the planning, implementation and evaluation of educational activities aimed at resolving an environmental issue that they themselves have identified.
- *International Case Studies on Water Quality Education* provides a rich picture of the kaleidoscope of programs world-wide. These case studies sensitize the reader to potential implementation barriers and offer a vast number of new ideas and resources for school and community based programs. (available late 1996)
- *Heavy Metals Manual* provides background information about heavy metals—ecological effects, sources or metals, policies, and laws. It also provides instructions for bioassay monitoring, instrument techniques and safety procedures. (available late 1996)

About GREEN

The impetus for GREEN began the spring of 1984 with a group of concerned students at a high school located along the polluted Huron River in Ann Arbor, Michigan. Their teacher contacted Dr. William Stapp and other educators at the University of Michigan, and together they developed a comprehensive educational program called GREEN.

The idea quickly caught on; experiences gained in three years of work with schools along the Huron set the stage for an expanded program on the Rouge River in 1987—part of an effort to improve education and the environment in the broader Detroit metropolitan area.

The educational model moved to other watersheds around the Great Lakes in the U.S. and Canada. As the program expanded nationally and then internationally, other components were added: community partnerships, computer telecommunications, cross cultural opportunities, and integration of GREEN's initiatives across the curriculum to form a comprehensive program for watershed sustainability.

Join the GREEN Network

We invite you to become a member of the Global Rivers Environmental Education Network, a community of global citizens dedicated to watershed stewardship and the enhancement of education.

➤ **Student Membership** ($5 per year) makes it possible for GREEN to offer you access to the GREEN World Wide Web home page and a GREEN KIDS membership card.

➤ **Supporting Individual Membership** ($25 per year) makes it possible for GREEN to offer you the quarterly GREEN Newsletter and access to a special rate GREEN/EcoNet Account.

➤ **Contributing Individual Membership** ($50 per year) makes it possible for GREEN to offer you the quarterly GREEN Newsletter, access to a special rate GREEN/EcoNet account, a 10% discount on items in the GREEN Catalog, and a 10% discount on Institutes and Workshops.

➤ **Sustaining Individual Membership** ($100 per year) makes it possible for GREEN to offer you the quarterly GREEN Newsletter, access to a special rate GREEN/EcoNet account, a 15% discount on items in the GREEN catalog, and a 15% discount on Institutes and Workshops.

➤ **Group Membership** ($500 per year and higher, depending on group size) makes it possible for GREEN to offer all group members the quarterly GREEN Newsletter, access to special rate GREEN/EcoNet accounts, a 15% discount on items in the GREEN Catalog, and a 15% discount on Institutes and Workshops. Minimum group size 25; please contact GREEN for details.

You can contact GREEN at:

GREEN
206 South Fifth Avenue, Suite 150
Ann Arbor, MI 48104
USA

Tel: (313) 761-8142
Fax: (313) 761-4951

Internet: green@green.org
WWW: <http://www.igc.apc.org/green>

I'd Like to Join GREEN

☐ Student Membership $5/yr
☐ Supporting Individual $25/yr
☐ Contributing Individual $50/yr
☐ Sustaining Individual $100/yr

William B. Stapp Endowment Fund

We have initiated a capital campaign honoring **Professor William B. Stapp,** founder and Honorary Director of GREEN. Dr. Stapp is recognized as one of the influential figures in modern environmental education. He served as Director of UNESCO's Environmental Education Program, and as president of the North American Association for Environmental Education. Dr. Stapp is a Nobel Prize nominee and Professor Emeritus at the University of Michigan School of Natural Resources and Environment.

I would like to make a contribution to the **William B. Stapp Endowment Fund.** I have enclosed payment in the amount of $ _____ .

Total Enclosed $ _____

☐ Check Enclosed
☐ Purchase Order # _____
☐ Tax Exemption # (if applicable) _____
☐ Credit Card # _____
  MC   /   Visa (circle one)
  Expiration Date: _____

Signature: _____

Print Name: _____

Address: _____

Tel: _____

Fax: _____

E-mail: _____

Contributions for Membership and to the William B. Stapp Endowment Fund
are tax deductible to the extent allowable by law.